Acoustic Essentials *for* ARCHITECTS

Michael Fay

ACOUSTIC ESSENTIALS FOR ARCHITECTS

J.ROSS PUBLISHING

ISBN: 978-1-60427-211-6
e-ISBN: 978-1-60427-866-8

Printed and bound in the U.S.A. Printed on acid-free paper.

10 9 8 7 6 5 4 3 2 1

Library of Congress Cataloging-in-Publication Data can be found in the WAV section of the publisher's website at www.jrosspub.com/wav.

Direct all inquiries to J. Ross Publishing, Inc., 151 N. Nob Hill Rd., Suite 476, Plantation, FL 33324. For EU safety/GPSR concerns, email: gpsr@mare-nostrum.co.uk or contact: Mare Nostrum Group B.V., Mauritskade 21D, 1091 GC Amsterdam, The Netherlands.

Phone: (954) 727-9333
Fax: (561) 892-0700
Web: www.jrosspub.com

To my wife Elaine, daughter Meredith, and brother Lew.
Without their love and steadfast support, this book would not exist.

CONTENTS

FOREWORD

Acoustics is architecture!

You have no idea how much I love seeing this phrase in print. As you will learn early in this book, this three-word phrase is attributed to my mentor in acoustics, architect and acoustician Russell Johnson. Russell was trying to express his frustration and change the way that acoustics were (or were not) being prioritized in the design of *sound-critical* spaces, such as concert halls and theaters.

In the mid-twentieth century, architects would lay out the design of sound-critical spaces and then engage an acoustical consultant to "do acoustics to the facility." This limited the ability of the acoustical consultants to influence key issues, such as room-shaping and other elements, that impacted the perceived acoustics of the space and defined the acoustics of the facility under design.

For many architects, acoustics was considered one of the *dark arts*, an unfathomable morass of spells and incantations suitable for the yet-to-be-written Harry Potter novels. Russell rejected this design method as fundamentally flawed. He dedicated his career to the idea that (as you will read early in this text), "You do *not* do acoustics to a room. Rather, you design a room to have the acoustics it needs."

Today, more and more architects are adopting this *design-in-the-acoustics* approach. "Why is this?" you may ask. Well, as you may have heard, the older method was not working so well, with many halls designed and built during the 1950s through the 1980s not living up to expectations for quality of sound. During this time there were also some spectacular failures, where rooms were opened and then had to be tweaked, twisted, renovated, and re-renovated to achieve even barely acceptable acoustics. A change in the design process was needed.

Since architects are the primary audience for this book, it is important to note that there are other people who play a huge role in prioritizing the focus of the architect's design efforts, including performers, audiences, production professionals, systems integrators, and facility owners. These groups are linked together by sharing an ever-increasing expectation for sound quality in every facility, be it new or a renovation.

During the last 70 or more years, work in acoustics has been driven forward by the military (~1950s through the 1970s), the automotive industry (1960s through the 1980s), the cinema industry (1970s through the 1990s), the music industry (1990s through the 2010s), and, of course, the telecommunications industry. Telecom has been the steady proponent for sound quality from the time of Alexander Graham Bell through today. A result of this wide range of industries working in acoustics is that the spaces we define as sound-critical have grown to include offices, conference rooms, teleconferencing centers, performing arts centers, home theaters, cars, cinemas, worship spaces, theaters, lecture halls, restaurants, opera houses, concert venues, corporate C-suites, and more.

All this brings us to Michael Fay, and his reference guide, *Acoustic Essentials for Architects.* As a pro audio and acoustical consultant who has worked on a wide range of projects, Michael also recognized early on that acoustics is not something that can be easily or effectively done to your room after the basic design has been defined. The architect needs to be engaged from the beginning of the design process.

Often, in order to create the space you want visually—and one that meets your client's needs—you need to lead (or at least be an active participant) in the discussion about acoustics. And to do that, you need to understand the essential topics included in this book.

So, this is a guidebook—similar to a well-valued travel guide. Michael has compiled a set of essential topics for architectural acoustics, presented in a clear, outlined format. These are not deep dives into the esoterica of acoustics, but rather, they are short, clear explanations of topics that come up when you are working through the design of a sound-critical space. The topics addressed include: what sound is; sound propagation; an acoustical lexicon of key words and phrases; echo, reverberation, and resonance; room geometry; noise; computer modeling; human hearing; speech intelligibility; variable acoustics; modern materials; applied acoustics; and many more.

This book explains and demystifies much of the acronym-filled world of acoustics. He has compiled a set of 21 short chapters, each providing useful and practical insights into what he thinks are the essential acoustical topics for members of the architectural and design community. As a working acoustical consultant, I plan to make this book a gift for every architect I am currently working with and, perhaps more important, for every architect to whom I have spent hours explaining and re-explaining many of the topics included here! As this book reaches a wider and wider audience, no longer will we in acoustics have to explain that *acoustic isolation* and *acoustical treatment* are two different topics.

In closing, it is important to note that just as audience expectations for sound quality are ever-increasing, the architectural and electronic technology and materials you are able to use to create the correct acoustical environments are also ever-changing. So, I will not be surprised if I hear that Michael has interrupted his retirement at some point in the future to update this primer. But fear not, most of the topics included here are foundational and will help you work more effectively today and into the foreseeable future—hopefully, with a cooperative and constructive acoustical consultant at the table by your side.

HAVE FUN!

—Sam Berkow, Founder, SIA Acoustics

Sam Berkow is an Acoustical Consultant/Designer based in NYC. He has worked on sound-critical spaces, including concert halls and production spaces, longer than he cares to admit. Mr. Berkow started his career working under the tutelage of architect and acoustical designer Russell Johnson. In addition to his work in facility design, Mr. Berkow is the developer of the Smaart™ Acoustical Measurement system, a leading acoustical measurement tool for over 30 years. In 2022, Mr. Berkow was awarded the Parnelli Lifetime Achievement Award for Innovation in Audio and Acoustics.

INTRODUCTION

I love working with open-minded architects. This book is inspired by the many architects I have had the privilege to work with during my long career as a pro audio and acoustical design engineer.

As both a *sound* and *acoustics* guy, I often felt and experienced an acoustical disconnect between those who design buildings and those who must live and work in them. During countless interactions, I realized how many architects are unknowing or unclear about how their designs impact a room's acoustics.

I tried sharing some of the basic concepts, but realized there was so much more for them to know and understand than what a few brief conversations could unveil. I then thought I could find and share some reference materials, but soon realized the absence of any suitable content. Everything was geared toward those practicing or studying to become acousticians, but nothing addressed the big picture, low-tech needs of the architectural community.

In 2010, I began writing the first edition of this tutorial as a white paper. While it was shared, presented at the 166th meeting of the Acoustical Society of America conference in 2013, and embraced by a few, it lacked the scope and graphic support needed for further publication. In 2022, I felt compelled to revisit this material with renewed enthusiasm.

One distinguishing feature that can be found in every room is its acoustical characteristics. Unfortunately, this thing we call *acoustics* is not well-known, understood, or even covered in most architectural schools (see Appendix A). Because of that fact, this reference guide is tailored to the one group of people having more influence over a building's acoustics than any other—*architects*.

Underpinning the core of many architectural projects is a foundation. The foundation must be strong enough to support all structural, functional, and creative ideas the owner and architect may need or want to consider. Yet, far too often, an acoustical foundation is not defined, approved, or implemented. Ultimately, architects are the drivers and coordinators of all construction decisions being made—somewhat like the *conductors* of an orchestra consisting of the owner(s) and many diverse consultants and contractors.

Formally, the subject being covered is called *architectural acoustics*. This is a field that is much deeper than most people imagine. To do justice to the theme, we must briefly touch on many subordinate topics, most having a synergetic relationship bonding architecture and sound.

This commentary is based on fundamentals and my 45+ years of experience—not perfection. It covers most of the basics, while also exploring several modern and esoteric matters. You will be introduced to interesting and analytical subjects; some you may know, some you may never have considered or understood.

Others may also find this reference enlightening. For example, most audio practitioners have to work in acoustical environments that are out of their control because of past decisions made by an architect, interior designer, and/or owner. Also, there are probably many new and veteran sound techs that have yet to accrue a solid acoustical foundation that allows them to understand what and why their sound is behaving well or badly. Most certainly, there are knowledge gaps on both sides of the designer/user relationship.

When opportunities arise, this commentary will help almost anyone better understand, discuss, and apply good acoustical principles.

EXAMPLES OF WHAT YOU WILL FIND IN THIS BOOK

- What is sound, and why is it so hard to manage or control?
- The length of commonly heard, low- and high-frequency sound waves vary by as much as 400:1. Why does this disparity matter?
- How and why do various audible frequencies behave differently when interacting with various materials, structures, shapes, and finishes?

- There are three acoustical tools available to both the architect and acoustician. What are they? How can they benefit or hinder the work of each craft?
- What is the wave/ray duality of sound?
- Is pure random accident part of your business model?
- Large rooms versus small rooms: How and why do acoustical challenges change as rooms become larger or smaller?
- Room geometry: Why some shapes are much better than others. Examples and explanations.
- Reverberation and echo: How do they differ? Which is better or worse and why? How much reverberation is too much or too little?
- New reverberation design goals for twenty-first century performance venues.
- What is the Parametric Method of Acoustic Treatment?
- Speech intelligibility: We all know it matters. What can architects do to help or hinder?
- Three simple tests: Quick, easy ways to evaluate the basic acoustical merits of a room without any fancy test equipment or training.
- Opportunities and trade-offs: Blending architecture, acoustics, and pragmatism.
- Applied acoustics: The four key feasibility questions.

THIS BOOK HAS FOUR MAIN GOALS

First—It will endeavor to put the essential terminology and concepts of architectural acoustics into the hands of the architect, owner, interior designer, end user, or anyone else who would like to have a better understanding of this subject.

Second—It will introduce the incredible dimensional range of wavelengths that exist between the lowest and highest audible sounds. Once defined and established, we can overlay this spatial understanding of sound waves onto the physical structures they touch.

Third—It will help architects better understand and visualize the relationship between the physical shapes, textures, and dimensions of their work, and why those design decisions impact the qualitative behavior of sound. For example, architecture requires three-dimensional thinking and implementation, yet sound is a four-dimensional experience that must co-exist within a three-dimensional structure. The fourth dimension, time, being significant because the speed of sound is so slow that the size and shape of a built environment can often dictate the quality and clarity of all sounds within.

Fourth—It will explain the importance of interior seating symmetry for medium and large rooms. Sound propagation is an amazingly flexible phenomenon but not something that is easily controlled once a group of sound waves begin to move in any particular direction. As a result, seating symmetry is an extremely important component related to achieving high-quality dispersion and reception of sound. Conversely, in small rooms, overall structural asymmetry becomes a priority.

It is easy to think that sound is completely removed from the fundamental principles of architecture. Nothing could be further from reality. Consider this: If a person were seated in an amazingly beautiful structure at night—one without illumination—they would see few, if any, of the features, shapes, materials, colors, textures, or workmanship that was created. However, given that same dearth of light, all audible *sounds* remain unaltered . . . for better or worse.

Note: Each chapter is presented in outline form to facilitate the numerous cross references being tossed between the various chapters and sections.

ACKNOWLEDGMENTS

Peer Review Team: A special thank you goes out to Neil Shade, FASA (Acoustical Design Collaborative); Charlie Hughes (Biamp and Excelsior Audio Labs); and Rob Miller (Threshold Acoustics) for assisting with notes, comments, and corrections that made this work presentable.

To my publisher/editor Stephen Buda at J. Ross Publishing: Thank you for recognizing the value of this material and believing that there is a market for such information.

Cover design and various figure graphics by Atiq Ahmed. www.atiqahmed.us.

In first name alphabetical order, others who have notably impacted or influenced my career and this work:

Aaron Ruse, Andreas Melcher, Bernd Noack, Bob Delson, Bruce Olson, Cam Bishop, Carl Yanchar, Dan Nelson, Dave Brown, Dave Paviol, Dave Shadoan, David McTeir, David Sheirman, Dennis Regan, Doug Kvandal, Eberhard Sengpiel, Everette Lybolt, Garrison Parkin, Geoff Williams, George Lampos, Glen Aubrey, Gordon Moore, Hadi Sumoro, Jamie Anderson, Jason Schwartzel, Jeff Schmitz, Jim Pilaar, Jim Plank, Joe Godfrey, John Storyk, Keith Clark, Kelly Fair, Larry McNamer, Lee Carrol, Marcus Jones, Marie Wright, Michael Lawrence, Mike Kettering, Mike Martin, Mike Plank, Mike Trimble, Mike Wofford, Neil Fink, Neils Adelman-Larsen, Pat Brown, Paul Tydelski, Peter D'Antonio, Peter Hester, Phil Leerar, Ralph Heinz, Randy Willis, Richard Lenz, Rick Stanley, Rob Grubb, Ron Suaro, Ross Ritto, Ryan Ash, Sameh Helmy, Scott Coyle, Steve Ellison, Steve Giffin, Steve Liddle, Steve O'Conner, Tyler Adams, and probably a few more that I have unintentionally omitted.

Architects who have notably impacted or influenced this commentary:

Bennett Lord—Lord Architecture

Bill Magnusen—Architects Mosher Drew

Bret Thoeny—BOTO Design Architects
David Keitel—Domusstudio Architecture
David Pfeifer—Domusstudio Architecture
John Groth—Groth Architects
Manuel Oncina—Manuel Oncina Architects
Mike Sebbins—CO Architects
Stan Andrade—Andrade Architects
Wayne Holten—Domusstudio Architecture

ABOUT THE AUTHOR

Michael Fay is owner/principal at GraceNote Design Studio, a pro audio and acoustic design consultancy. He is also a member of SynAudCon, the Audio Engineering Society (AES), the Acoustical Society of America (ASA), AVIXA, the Miramar College Audio Program Advisory Board, and former member of the Art Institute Audio Program Advisory Board.

Mike's experience includes more than 45 years of professional AV and acoustical systems design, systems engineering, recording studio engineering, live audio mixing, and technical writing. Now semi-retired, he spent the last 25 years of his full-time career as a senior-level systems design engineer, executive project manager, technical integration manager, technical sales engineer, and acoustician.

His pro audio experience sprung from a foundation of classical music training and performance (trumpet) in his youth and continued to develop in college with an emphasis on music theory, broadcasting, and electronics. From an early age, he was drawn to the quality (or lack thereof) of the recordings he listened to. So much so, that at the age of about 12 he was convinced he could do a better job mixing most of the pop records he heard. At that time his favorite recordings were motion picture soundtracks.

His passion for music and sound eventually led him into the recording studio where he worked primarily on corporate/industrial recordings for TV, radio, film, and theme park shows. Within just a few years he was considered the first call independent studio engineer in San Diego. While working in some of the top studios in Southern California, he also developed a fascination for studio design and acoustics, which ultimately led him into the field and study of architectural acoustics.

In 1987, Mike became the editor of Recording Engineer/Producer (RE/P) magazine, a highly regarded, internationally distributed, technical trade magazine servicing both the recording and touring sound industries. In the early 1990s, he moved back into the applications side of pro audio, making the transition into systems design and integration.

His work history includes time as: Owner of Michael Fay & Associates, Technical Director at College Avenue Baptist Church, Director of Engineering at Audio Associates, and Senior Design Engineer and General Manager of Sound Image's Integration Division, which was and continues to be listed as one of the country's top 50 AV integration firms.

In late 2017, he retired from full-time employment and launched Grace-Note Design Studio. From here, he continues to design, consult, write, and mentor the next generation of audio and acoustical professionals, architects, owners, and end users. For the past three and a half decades, he has considered his deep knowledge, understanding, appreciation, and commitment to music and sound to be his *spiritual gift* from God.

For more information, Mike can be reached via his website: www.gracenoteds.com.

Web Added Value™

This book has free material available for download from the Web Added Value™ resource center at *www.jrosspub.com*

At J. Ross Publishing we are committed to providing today's professional with practical, hands-on tools that enhance the learning experience and give readers an opportunity to apply what they have learned. That is why we offer free ancillary materials available for download on this book and all participating Web Added Value™ publications. These online resources may include interactive versions of material that appears in the book or supplemental templates, worksheets, models, plans, case studies, proposals, spreadsheets and assessment tools, among other things. Whenever you see the WAV™ symbol in any of our publications, it means bonus materials accompany the book and are available from the Web Added Value Download Resource Center at www.jrosspub.com.

Downloads for *Acoustic Essentials for Architects* include the T_{60} Slope Ratio Thesis (New Reverberation Design Goals for Modern Architectural Acoustic Environments), Parametric Acoustics, Second Edition (The Case for the Parametric Method of Acoustic Treatment), and Acoustic Shadowing (Unpacking Wave Numbers and Introducing the Wave Ratio Thesis).

CHAPTER 1

ARCHITECTURAL ACOUSTICS DEFINED

1.1 Architectural acoustics (AA) can be defined as the science, study, and application of acoustical principles as they are implemented inside a building or structure. In the context of this book, the terminology focuses on built environments that will be used for live, amplified, creative performances, or the audible presentation of other useful information. In other words, medium and large room acoustics. To a lesser extent, small room acoustics will also be covered.

1.2 As defined by the architecture dictionary, "Acoustic design refers to the strategic planning and implementation of elements within a building to achieve optimal sound quality and control. This involves managing how sound is generated, transmitted, and received within a space to enhance clarity, reduce unwanted noise, and create an acoustically pleasant environment."[1]

1.3 Architectural acoustics and sound have an interdependent relationship. Once established by design, right or wrong, they are inseparable. They coexist as a bidirectional cause-and-effect engine.

1.4 While AA includes topics such as interior noise control and managing sound transmission in and out of a room or building, the focus of this reference is managing reverberation and resonances that directly affect the quality and clarity of sound in any venue.

1. https://www.architecture.org/learn/resources/architecture-dictionary/entry/acoustic-design/

A. Too much reverberation causes speech and music clarity to suffer greatly. If there is too little reverberation, things like acoustic instruments, musical vocals, and congregational singing are dull, lifeless, or unblended—like they often are when performing outdoors.

B. Room resonance is often problematic at the lower end of the frequency spectrum, but can rise into the low to mid speech frequencies in small rooms.

C. Reverberation and resonance are covered in much more detail in Chapter 5.

1.5 Intentional or not, acoustics are a crucial byproduct of architectural and interior design. The product of those disciplines significantly influences our auditory experiences in all spaces. "By striking a balance between aesthetics and functionality, architects and designers can shape architectural spaces that engage our senses and enhance our connection to sound. From concert halls to classrooms, offices to residences, the art and science of acoustics continues to shape the way we experience the world of sound."[2]

1.6 As legendary architect and acoustician Russell Johnson[3] often said, "You do *not* do 'acoustics' to a room. Rather you design a room to have the acoustics you need!" Refined even further, he proclaimed, "Acoustics is architecture!"

2. https://www.easternengineeringgroup.com/acoustics-101-understanding-the-basics-of-sound-control-in-architectural-spaces/

3. https://en.wikipedia.org/wiki/Russell_Johnson_%28acoustician%29

CHAPTER 2

WHAT IS SOUND?

2.1 Have you ever stopped to consider just what sound is? Of our five senses, hearing seems to be the one receiving the least consideration, though some modern research[1] suggests it's the most important for our overall well-being and lifestyle. Why then do our listening environments garner so little attention and respect? The answer is at least threefold.

A. Sound and acoustics are relatively complex subjects that are poorly understood by most people.

B. Sound systems and acoustical treatments must be customized for each room. There is no such thing as a *one-size-fits-all* solution. To reach a room's desired audio quality and clarity, different sizes, shapes, locations, and quantities of loudspeakers and acoustical finishes are often required.

C. Quality loudspeakers and acoustical treatments are often perceived as being expensive, bulky, and/or generally unsightly. While historically true, the size and appearance of these products are becoming more aesthetically appealing.

2.2 Sound is not a tangible thing; it is a transformative experience.

A. Sound is a relatively slow-motion, push/pull, molecular chain reaction that starts with a simple oscillation or vibration. These vibrations are created by any number of humans, animals, or objects.

---

1. https://www.acousticbulletin.com/our-visual-focus-part-2-the-eye-versus-the-ear

B. The vibrations become a sound we can hear if they stimulate the nearby air molecules with enough energy to be detected. Given a very early age and optimal conditions, the audible spectrum for humans is generally considered to be 20 Hertz (Hz) (named after German physicist Heinrich Hertz) to 20,000 Hz.

C. As any object vibrates, there is an energy transfer between adjacent air molecules that corresponds to the frequency of its vibrations. To visualize this energy transfer, think about Newton's cradle (see Figure 2.1) of steel balls. Regardless of how many, the balls in the middle really don't move. Only the first and last can be easily seen to move because they are the beginning and end of a chain reaction. At the molecular level, the electrons in air squeeze together just enough to transfer their *new-found* energy along, then return to their original position—via compressions and rarefactions in the air

Figure 2.1 The Newton cradle is an excellent visual example of the energy transfer between energized and static air molecules. Said another way, the conversion of potential energy to kinetic energy.

molecules—thereby creating small areas of higher and lower air pressure (see Figure 2.2).

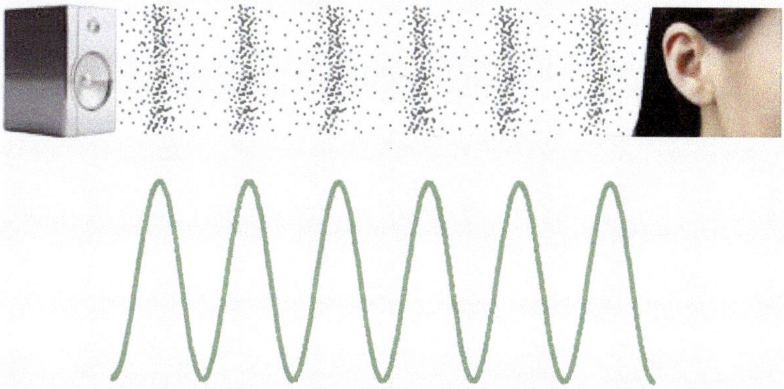

Figure 2.2 This graphic shows an overly exaggerated view of a sound wave moving through the air—from source to ear. Graphic by Atiq Ahmed

1. The stronger or more violent the oscillating motion, the greater the change in pressure. This results in a louder version of the vibration or sound.

D. Barring obstruction, these tiny changes in air pressure move away from the source of the vibrations and travel through the air rather slowly, until they strike another set of objects that sympathetically vibrate. For us humans, those sympathetically vibrating objects are our eardrums.

E. In its most rudimentary form, the transformative circuit that culminates in audible sound looks like this: A vibrating object → a chain reaction of air molecules pushing toward and pulling away from one another → one or more sympathetically vibrating eardrums → brain → *qualia*[2]; the subjective human experience. Individual air molecules make no audible sound.

2.3 Sound can also be thought of as a force, somewhat like gravity or magnetism. But unlike gravity with its ubiquitous downward force or magnetism with its *sticky* bipolar forces, think of sound as a force that generates molecular chain reactions. When man-made and intentional, it is a force that can easily be activated or deactivated, when and if required.

A. Intentional sound waves can easily move through air in any direction; they can be manipulated and steered into relatively definable areas; and with proper care and handling are generally adaptable and responsive to our needs.

B. However, sound waves cannot be bent around corners or penetrate solid objects without significant quality, quantity, and/or temporal changes; usually unwanted.

1. When sound waves bend around objects or corners, it is called diffraction.

2. When sound waves strike solid objects, some of that energy may be transferred via structure-borne vibrations, which can plague noise isolation efforts out of or into a room.

2.4 There must be some type of *compressible medium* for sound to exist. It doesn't matter what type of medium or atmosphere, so long as there are molecules that can be pushed around or stimulated into distinct patterns of greater and lesser pressure.

A. Water is a good example of a medium that easily transmits sound; so too are steel and concrete, under the right conditions.

B. A vacuum has no atmosphere; therefore, sound does not exist in a vacuum.

---

2. https://en.wikipedia.org/wiki/Qualia

2.5 Frequency, wavelength, and the speed of sound:

A. Sounds can be defined by either their *frequency* or *wavelength*. Every discrete frequency has a corresponding wavelength. While frequency and wavelength are directly correlated and interchangeable, each word defines something different.

1. When identifying a specific frequency in cycles per second, the unit of measurement is Hz.

2. To say that a sound has a frequency of 500 Hz means something is vibrating, back and forth, 500 times each second.

3. As any discrete vibrating frequency is produced and propagates or moves outward at the speed of sound, the linear distance required to complete one full, sinusoidal wave or cycle is its wavelength. Thus, the wavelength of a pure 500 Hz tone is about 2.25′ and a 2.25′ wave is a 500 Hz tone. They are identical in size and sound.

B. When traveling through air, the speed of sound is a *generalized* constant. The number used here, and referenced throughout, is 1,127 feet per second (FPS). This number will fluctuate slightly based on temperature and humidity, but not enough to be meaningful in the context of this tutorial.

1. The complex waves of music and speech all travel through air at the same speed, regardless of frequency.

2. If you're interested, the math conversion from frequency to wavelength is very simple: divide the speed of sound (1,127) by whatever frequency you are interested in. For example: 1,127 FPS ÷ 320 Hz = 3.52′.

C. Now compare the speed of sound, at 1,127 FPS, to the speed of light, which is about 186,282 miles per second.

1. Because the wavelengths of visible light are so small, directing, shaping, and containing light is many orders

of magnitude easier than shaping, containing, and controlling sound. Why? Because the longer the wavelength, the more difficult it is to control, dissipate, and/or stop once it is in motion.

2. It is relatively easy to completely block or limit the visibility of any unwanted light. One way is to simply close a door between two rooms. If needed, and with little effort or specialized planning, the light in room *A* can be completely eliminated from room *B*.

3. This same exercise is much less effective if you want to block or limit some or all audible sounds between the two rooms. Low and mid frequency sounds (LF and MF) can easily penetrate doors, walls, and windows, usually via the transference of structural vibrations. Managing speech privacy between two adjacent rooms is a perfect example of such challenges.

4. Only with very careful structural planning and/or specialized acoustical treatments can we completely stop and/or block all sound.

D. Understanding how audible frequencies relate to wavelengths is a vital link between sound, acoustics, and architecture. Once sound is translated into physical dimensions that architects or interior designers can easily understand, they can begin to unlock the mystery of why effective acoustical treatment needs to be applied in so many sizes and shapes.

CHAPTER 3

SOUND PROPAGATION

3.1 Desired sound, not random noise, can be propagated or dispersed with some reasonable control. This is primarily done via the use of specific loudspeaker types and technologies.

A. Once again, the wavelength of various sounds largely limits how much control is available. In the lower frequency spectrum, sound emanating from instruments and/or loudspeakers is generally omnidirectional. As frequencies increase in pitch, wavelengths get smaller and smaller, and more directional control can be gained.

B. 500 Hertz (Hz) is a good point of reference. Amplified sound—at or above 500 Hz (2.25′ wavelength)—can often be directionably controlled. Controlling the direction of sound waves becomes more and more difficult for frequencies below about 500 Hz. Difficult, but not impossible.

1. For a quick reference, 500 Hz is fairly close to the lowest tone that most people can whistle.

3.2 Sound wave modeling:

A. Sound travels through the air in longitudinal waves. While these waves may not be too difficult to visualize at the most basic level, their reflective behavior can be quite difficult to accurately calculate and predict in three-dimensional space.

B. Architectural acoustics correlates the myriad frequencies and wavelengths of audible sound, with the complexities of architectural geometry, and the influences of multiple finish materials. Together, these factors are very difficult to compute using manual methods.

C. Today, we rely heavily on computer-based 3-D modeling, calculations, and simulations to help show (see Figures 3.1, 3.2, and 3.3) and solve the most critical and complex acoustical questions.

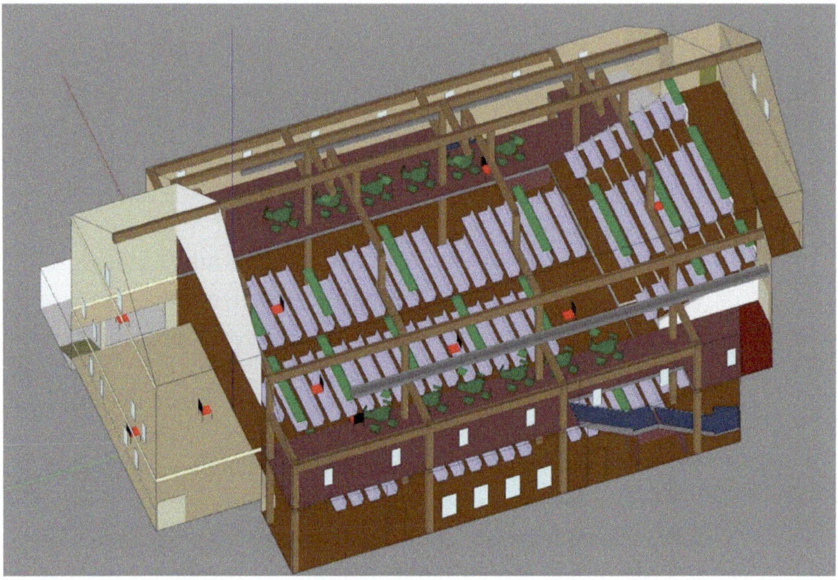

Figure 3.1 Rendering of a 3-D acoustical room model. Modeling done in EASE 4.0 electro-acoustic software application.

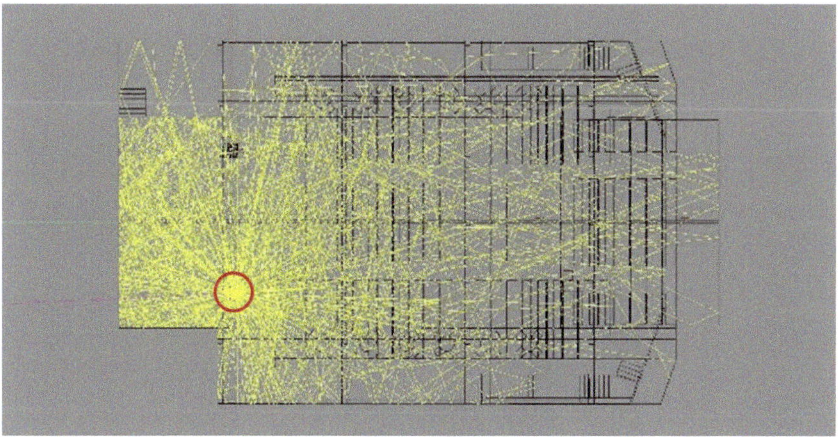

Figure 3.2 A wireframe plan view of the room in Figure 3.1 with 1,000 Hz ray tracing beams in yellow. The red circle indicates where an omnidirectional source loudspeaker is located. The beams shown represent the direct path and the first- and second-order reflections.

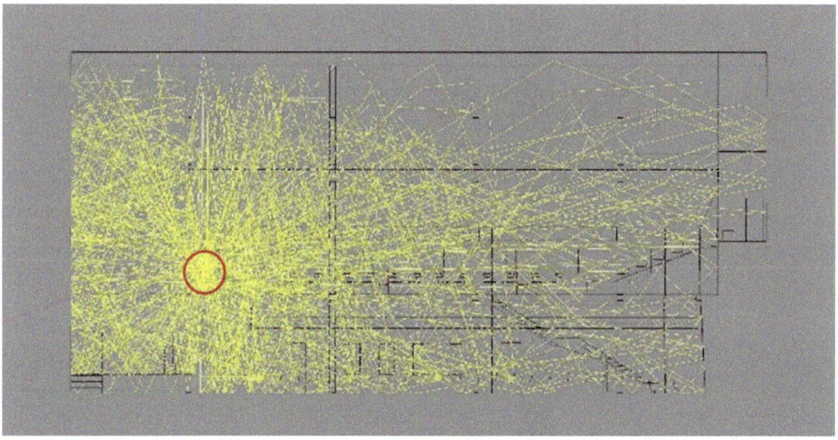

Figure 3.3 A wireframe elevation view of the Figure 3.2 beams.

D. *Ray tracing* theory and software algorithms are used to model the behavior of sound as it radiates from a virtual sound source and bounces around in a fully enclosed 3-D computer model. The virtual rays used in modeling programs can be thought of as discrete, laser-like beams of light. Imagine a starburst of light beams, with each beam expanding out from a virtual source, like a large fireworks artillery shell bursting in the night sky.

   1. Most modeling programs emit these beams at a spherical density of between one and five degrees of separation between each beam.

E. Image casting: The most advanced computer modeling software applications incorporate *image casting*, which provides much faster calculations. Image casting is a recent variation on ray tracing methodology. Instead of simulating individual rays, image casting creates virtual *images* of the sound source at each reflective surface. These virtual images act as new sound sources, simplifying the calculations.

F. Both ray tracing and image casting can effectively approximate the behavior of direct and reflected sound as it travels through air. But there's a catch:

   1. Ray tracing and image casting are best used to represent only a portion of the audible frequency spectrum. More specifically, only about six of the 10 total octaves (see Section G) of audible sound (those between 320 Hz and 20 kilohertz (kHz) can be modeled with reasonable accuracy.

      a. When using either technique, accurately modeling frequencies between 20 Hz and 320 Hz is questionable, primarily because the reflected sound energy is moving into the modal (room resonance) frequency regions. The modeling software becomes less and less accurate and reliable. See Chapter 5, Section 5.5 for more on resonance and room modes.

b. For reference, those wavelengths are 56.35′ (this is not a typo) for 20 Hz, and 3.52′ for 320 Hz.

G. An *octave* is a musical term. All musical notes have a correlated fundamental frequency and wavelength. When any musical note is halved or doubled, its corresponding frequency and wavelength are halved or doubled. These are octave intervals (see Figure 3.4).

1. An octave below A440 is A220. An octave above is A880. The numbers represent the fundamental frequency of these three musical notes, and each has a corresponding wavelength that is doubled or halved in wavelength size.

a. The wavelength of A440 is 2.56′. A220 is 5.12′, and A880 is 1.28′. Across just these three octaves, the wavelengths vary by a factor of 4—between the highest and lowest note.

3.3 Delivery methods:

A. There are three primary, yet distinctly different, methods of transmitting the sounds (propagation) of a performance or presentation to a live audience. They are:

1. Acoustically, with no electronic reinforcement or amplification.

a. Relying on pure acoustic sound propagation is generally reserved for very sophisticated rooms and end users. Symphonic concert halls come to mind. These rooms require very strict design criteria and construction techniques, and though important, are not the primary focus of this commentary.

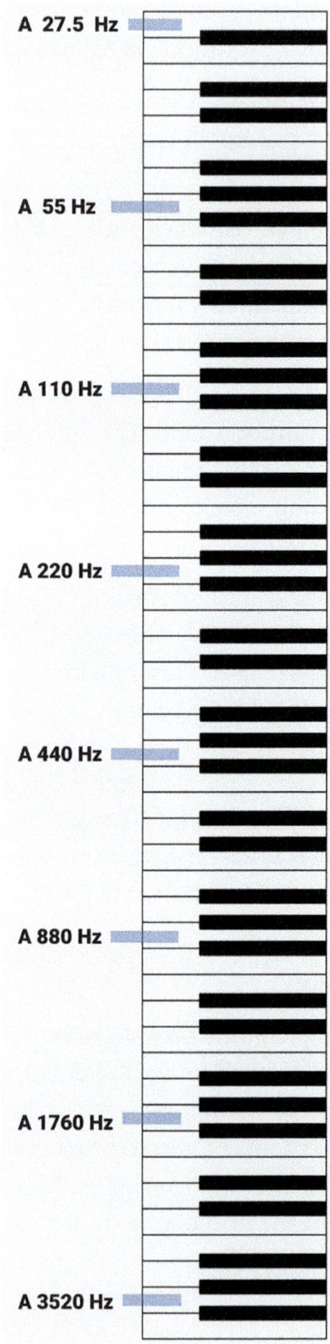

Figure 3.4 A piano keyboard that points out eight of its corresponding octave-related frequencies. Notice that each of the *A* notes is a doubling or halving of its closest neighboring *A* frequency.

2. Point source, or line array, sound reinforcement systems.
   a. This reference is written with these reinforcement systems clearly in mind, as they represent the most commonly specified technologies used in facilities that seat more than about 50 people.
   b. These are also the systems that can produce the widest frequency response, the greatest amount of power and volume, and that will generally benefit most from being installed in a good acoustical environment.
3. Overhead, distributed loudspeaker systems.
   a. These, too, are common systems but are typically specified to support background music, light foreground music, and/or speech-only sound reinforcement.
   b. These systems are often specified when more limited frequency response and/or power requirements are required.
   c. The overhead, distributed loudspeaker system is often the most effective approach when a low ($\leq 15'$) flat ceiling is specified.
   d. There are other delivery methods, but they usually operate independently of a room's architectural acoustics.

CHAPTER 4

ACOUSTICAL LEXICON

4.1 There are hundreds of unique words, phrases, and acronyms in the acoustical lexicon. It is not necessary for architects to know or understand most of them. However, there are a few that are essential and widely used, therefore the architectural and interior design (ID) communities should have a basic understanding of their meaning and applications.

A. Along with their acronyms, Greek letters, and symbols, the terms/phrases are: decibel (dB); sound pressure level (SPL); absorption coefficient (α or alpha); noise; noise reduction coefficient (NRC); speech transmission index (STI); articulation loss of consonants (%ALcons); sound transmission class (STC); noise criteria (NC); and delay (see Figure 4.1).

Units	Description
dB	Decibel
SPL	Sound Pressure Level
α or alpha	Absorption Coefficient
NRC	Noise Reduction Coefficient
STI	Speech Transmission Index
%ALcons	Articulation Loss of Consonants
STC	Sound Transmission Class
NC	Noise Criteria
delay	Time offset, typically noted in milliseconds (ms)

Figure 4.1 Commonly used acoustical terms, acronyms, Greek letters, and symbols.

B. Add echo, reverberation, and resonance to that list. Because these three are possibly the most important to grasp and understand, they are presented in more detail in Chapter 5.

C. It is not necessary to memorize the details. Just be aware that this terminology exists and be familiar with how it fits into most acoustical conversations.

4.2 The *decibel* (dB) is a relative unit of measurement that is widely used in audio and acoustics. The dB is a logarithmic ratio of the difference between two values, or a measured value and a reference (threshold) value of 0 dB.

4.3 *SPL* is the acronym for sound pressure level, measured in dB. The threshold of human hearing has a reference sound pressure of 0.00002 Pascals (Pa), which is equal to 0 dB SPL. This is the threshold of hearing in air at 1 kilohertz (kHz).[1]

A. Decibels are used to compress the immense range of Pa we can hear into manageable numbers. Pa is a unit of pressure (see Figure 4.2).

Sources @ 1 m	Sound Pressure	SPL re. 20 μPa
Rifle	200 Pa	140 dB
Threshold of pain	63.25 Pa	130 dB
Threshold of discomfort	20 Pa	120 dB
2 power mowers	2 Pa	100 dB
1 power mower	1 Pa	94 dB
Street traffic	0.2 Pa	80 dB
Talking	0.02 Pa	60 dB
Library	0.002 Pa	40 dB
Recording studio	0.0002 Pa	20 dB
Ref. sound pressure	0.00002 Pa	0 dB

Figure 4.2 This SPL table shows the huge range of Pa we may be exposed to and the more practical SPLs, in dBs.

1. Search Threshold of Hearing at https://www.acoustic-glossary.co.uk/index.htm

B. SPL is a sound field quantity and uses the 20 log factor; therefore, as a rule of thumb:

1. A 6 dB volume change is equal to a sound that is generally perceived to be about half again louder or softer (+/− 1.5x).

2. A 10 dB volume change is equal to a sound that is generally perceived to be about twice or half as loud (+/− 2x).

3. A 20 dB volume change is equal to a sound that is generally perceived to be about four times as loud or soft (+/− 4x).

4.4 The *absorption coefficient* (alpha or α) characterizes the fraction or percentage of sound energy absorbed—per square foot or square meter—by any given material. Expressed as a frequency-dependent value between 1.0 equals perfect absorption (no reflection) and 0 equals zero absorption (total reflection). The values for each material vary with frequency, angle of incidence, and mounting method. The alpha values for each material are determined experimentally.

A. One square foot of 100% sound-absorbing material has a value of 1 imperial sabin.

B. One square meter of 100% sound-absorbing material has a value of 1 metric sabin.

C. Note: The historical understanding that the absorption coefficient previously defined is limited to a maximum value of 1.0 is currently being challenged. Laboratory research[2] has been ongoing for about 10 years with a goal to establish and publish a new ISO standard. If successful, it would eventually replace the ubiquitous *absorption coefficient* with a more accurate *absorption constant*. The absorption constant will factor in the additional *edge effect* absorption values caused by the refraction of sound energy onto an absorptive surface.

2. http://nwaalabs.ipower.com/test.html

D. The *sabin* is a unit of sound absorption and an application subset of the absorption coefficient. When evaluating a comprehensive acoustical treatment plan, sabins are used to calculate the total absorption provided by all the various treatment materials being installed. Put into context: a 10 ft^2 piece of material, which has an alpha of 0.70 at 1 kHz, will provide 7.00 sabins of absorption at 1 kHz.

4.5 *Noise* is any undesired sound that is audible to the recipient. All sounds that are not productively contributing to the desired program content are noise. Examples include coughing, sneezing, crying, HVAC turbulence or whistles, electrical hums or buzzes, aircraft, highway, train, trolley, industrial, commercial, and/or other noticeable in-room activities.

A. Discrete echoes and excess reverberation are noise, too.

4.6 The *noise reduction coefficient* (NRC) is a single number value ranging from 0.0 to 1.0. NRC describes the average sound absorption performance of a material [across the sound spectrum associated with speech]. An NRC of 0.0 indicates the object does not attenuate mid-band sounds, but rather reflects sound energy. Conversely, an NRC of 1.0 indicates that the material provides 100% absorption [of mid-frequency sounds] for an acoustical surface area (in units sabin) that is equivalent to its physical, two-dimensional surface area.[3]

A. NRC is the average of the absorption coefficients for a specific material and mounting condition determined at the octave band center frequencies of 250, 500, 1,000, and 2,000 Hz.

B. Note: Because it averages a limited number of octaves, NRC is the least useful metric for any meaningful acoustical analysis, or treatment, involving music or other low-frequency (LF) content.

---

3. https://en.wikipedia.org/wiki/Noise_reduction_coefficient

4.7 Speech transmission index (STI) and articulation loss of consonants (%ALcons).

A. The understanding of speech (speech intelligibility) is directly dependent on the background noise level, the reverberation time, and the size of the room. The loss or confusion of the consonants contained in the words we hear leads to poor speech intelligibility. To rate or grade speech understanding we have different methods, where STI and %ALcons are established.[4]

1. STI is a machine measurement of intelligibility whose value varies from 0 (completely unintelligible) to 1 (perfectly intelligible). STI must be measured in situ and cannot be predictively calculated.

2. %ALcons expresses the *loss* of consonant definition, and can be predictively calculated during the design process. Lower scores are associated with greater intelligibility. Zero %ALcons indicates perfect clarity. For learning environments, a loss of 5% or less is desired.

B. STI and %ALcons basically communicate the same information, even though they are calculated and expressed differently. When properly applied, either is applicable (see Figure 4.3). See Chapter 10 for more on speech intelligibility.

STI	0 - 0.30	0.30 - 0.45	0.45 - 0.60	0.60 - 0.75	0.75 - 1.00
	Unacceptable	Poor	Fair	Good	Excellent
%Alcons	100 - 33%	33 - 15%	15% - 7%	7 - 3%	3 - 0%

Figure 4.3 This chart shows how STI and %ALcons arrive at the same mean opinion score grades, from unacceptable to excellent.

4. https://en.wikipedia.org/wiki/Sound_transmission_class

4.8 The *Sound Transmission Class* (STC) is an integer rating of how well a building partition attenuates airborne sound (see Figure 4.4). In the United States, it is widely used to rate interior partitions, ceilings, floors, doors, windows, and exterior wall configurations. Outside the United States, the ISO Sound Reduction Index (SRI) is used. The STC rating very roughly reflects the dB reduction of noise that a partition can provide (see Figure 4.5). The STC is useful for evaluating annoyance due to speech sounds, but not music or machinery noise since these sources contain more LF energy than speech.[4]

STCs by Audio Distinguishability	
STC	**What Can Be Heard**
25	Normal speech can be understood
30	Loud speech can be understood
35	Loud speech audible, but not intelligible
40	Loud speech audible as a murmur
45	Loud speech detected, but not audible
50	Loud sounds faintly heard
60+	Good soundproofing. Most sounds do not disturb

Figure 4.4 This chart should be used as a guideline to manage expectations. The higher the STC number in the left column (rated in dB), the more effective a wall assembly is at blocking speech range sounds from passing through. Tables like this are highly dependent on the background noise levels in the receiving room. Sound isolation is perceived to be better when the background noise in the receiving room increases.

STCs by partition type

STC	Partition type
27	Single pane glass window (typical value) (dual-pane glass window range is 26–32)"STC Ratings".
33	Single layer of ½-inch (13 mm) drywall on each side, wood studs, no insulation (typical interior wall)
39	Single layer of ½-inch (13 mm) drywall on each side, wood studs, fiberglass insulation[33]
44	4-inch (100 mm) hollow CMU (concrete masonry unit)[34]
45	Double layer of ½-inch (13 mm) drywall on each side, wood studs, batt insulation in wall
46	Single layer of ½-inch (13 mm) drywall, glued to 6-inch (150 mm) lightweight concrete block wall, painted both sides
46	6-inch (150 mm) hollow CMU (concrete masonry unit)[34]
48	8-inch (200 mm) hollow CMU (concrete masonry unit)[34]
50	10-inch (250 mm) hollow CMU (concrete masonry unit)[34]
52	8-inch (200 mm) hollow CMU (concrete masonry unit) with 2-inch (51 mm) Z-bars and ½-inch (13 mm) drywall on each side[35]
54	Single layer of ½-inch (13 mm) drywall, glued to 8-inch (200 mm) dense concrete block wall, painted both sides
54	8-inch (200 mm) hollow CMU (concrete masonry unit) with 1½-inch (38 mm) wood furring, 1½-inch fiberglass insulation and ½-inch (13 mm) drywall on each side[35]
55	Double layer of ½-inch (13 mm) drywall on each side, on staggered wood stud wall, batt insulation in wall
59	Double layer of ½-inch (13 mm) drywall on each side, on wood stud wall, resilient channels on one side, batt insulation
63	Double layer of ½-inch (13 mm) drywall on each side, on double wood/metal stud walls (spaced 1 inch [25 mm] apart), double batt insulation
64	8-inch (200 mm) hollow CMU (concrete masonry unit) with 3-inch (76 mm) steel studs, fiberglass insulation and ½-inch (13 mm) drywall on each side[35]
72	8-inch (200 mm) concrete block wall, painted, with ½-inch (13 mm) drywall on independent steel stud walls, each side, insulation in cavities

Figure 4.5 Examples of various STC ratings that may be achieved using various construction materials. The STC values represent sound attenuation in dB. *Source*: https://en.wikipedia.org/wiki/Sound_transmission_class

A. STC and NRC are different rating systems. They are easily and often confused by many.

1. The STC rating is used to evaluate the amount of sound that can be blocked by a material or assembly. Use this when evaluating partitions or walls meant to block sounds between rooms.

2. NRC ratings tell us how much mid-band sound is being absorbed by a given material. Use this when evaluating products meant to help improve speech intelligibility or reduce reverberation time in a room.

3. Achieving a desired STC or NRC score for one of these alone does not mean you will achieve a good result in the other; they are mutually exclusive.

4.9 *Noise criteria* (NC) uses a single number for rating the ambient noise levels in a room. It was developed in the United States in the 1950s, and is used extensively by the HVAC industry to test the background noise levels in offices, etc. The measured octave bands are compared with NC curves, which are based on the generalized average sensitivities for human hearing, aka the equal loudness contour. The NC rating is the value of the highest NC curve line touched by, but not crossing, the measured noise spectrum (see Figure 4.6).

A. Figure 4.7 shows examples of the generally recognized NC and dB_A targets for various common venues.

1. dB_A means that an *A-weighted* filter has been applied to instrument-measured sound levels (in dB) to account for the relative loudness perceived by the human ear. This is done because the ear is less sensitive to LF sounds.

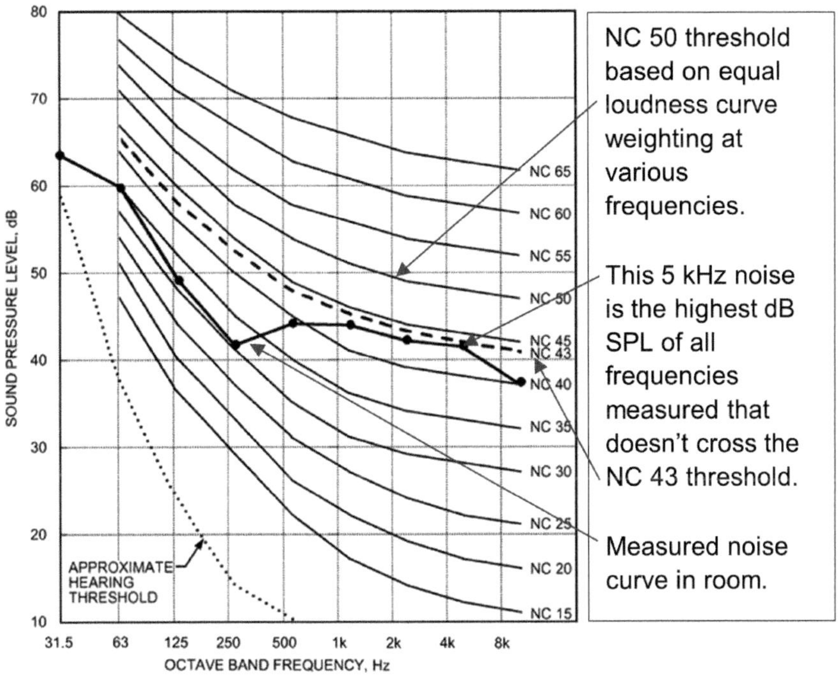

Figure 4.6 This NC chart shows a room with an NC 43 rating. The rating is determined by the measured ambient noise level at each of the frequencies shown at the bottom of the chart (with implied 1/3 octave resolution). Because human hearing is not linear, an approximation of the equal loudness contour[5] is laid under the measured noise. Whatever frequency touches highest at any given NC line, sets the NC rating for the whole room. In this measurement, some form of whine or whistle at about 5 kHz is the loudest noise, relative to the hearing sensitivities of our ears.

5. https://en.wikipedia.org/wiki/File:Lindos1.svg

Type of Room	Recommended NC Levels	Equivalent Sound Level dBA
Assemply Hall	25-30	35-40
Houses of Worship	30-35	40-45
Courtrooms	30-40	40-50
Private Homes	20-30	30-42
Meeting & Banquet Halls	25-35	35-45
Conference Rooms	25-30	35-40
Lecture & Classrooms	25-30	35-40
Motion Picture Theaters	30-35	40-45
Concert & Recital Halls	15-20	25-30
Indoor Sports Venues	45-55	55-65

Figure 4.7 Examples of recommended NC levels in various rooms and venues.

4.10 *Delay*: As referenced in Chapter 2, Section 2.5.B, the speed of sound is very slow, especially when compared to the speed of light. This is significant because unsynchronized sounds, that is, identical sounds that are separated by a delay of about 30 ms (milliseconds) or more are heard as separate sounds. This is problematic for many reasons. The longer the delay, the more noticeable and problematic it becomes.

A. For most people, the ear/brain system is able to integrate delays that are less than about 30 ms; therefore, very short delays (less than 30 ms) are generally indistinguishable from the original sounds and are usually beneficial.

B. A common example of problematic delay occurs when talking on a cell phone. Hearing your own voice come back to you—a fraction of a second or more after you have said something—is a nearly identical, delayed sound.

C. In medium and large rooms, a common example is caused when one or more point source loudspeakers are installed throughout a room to provide coverage in areas that the main loudspeakers cannot adequately reach. The need for such secondary *fill* loudspeakers is generally due to distance, obstructions, and/or for asymmetrical or atypical seating layouts. All fill loudspeakers need to be carefully time-aligned (synchronized) to eliminate any perceived delays. See more on this topic in Chapter 14.

   1. When necessary, and done properly, additional fill loudspeakers add significant cost to an audio system. Mirror image seating symmetry relative to the center line of the room—that is, equal in size, shape, finishes, elevation, and distance from the stage, platform, or desired sound source—generally minimizes the number of fill loudspeakers required.

D. Example of a worst-case scenario: Unsynchronized loudspeaker delays deployed in a room with unmanaged echo and excess reverberation. With reasonable accuracy, all these can be predetermined using the %ALcons calculations.

CHAPTER 5

ECHO, REVERBERATION, AND RESONANCE

5.1 The terms echo, reverberation, and resonance are commonly used, and often misused, in acoustical conversations and communications. In such discussions, it is important that everyone has the same basic understanding of these terms. They are not interchangeable.

5.2 Unmanaged echo, reverberation, and/or resonance are often the greatest acoustical challenges. For better or worse, they represent the net acoustical impact of owner, architectural, and/or interior design decisions.

5.3 Echo:

A. "Echo is heard as sound energy that has been reflected, and arrives with such a magnitude and time interval after the direct sound, as to be distinguishable as a repetition of it."[1] Echo has two different characteristics: *slap* and *flutter*.

B. Slap echoes are heard when sound waves are reflected back to their source or to a listener one time (see Figure 5.1).

C. Flutter echo occurs when sound waves bounce back and forth, multiple times, between hard, reflective, parallel walls, or other large surfaces. Regardless of room size, flutter echo is most commonly audible in purely rectangular rooms, with

1. https://www.acoustic-glossary.co.uk/index.htm

a moderately short (less than about 1.50 seconds) reverberation time.

D. Concrete, brick, wood, sheetrock, glass, tile, and steel are examples of finish materials that are hard, and nearly 100% reflective.

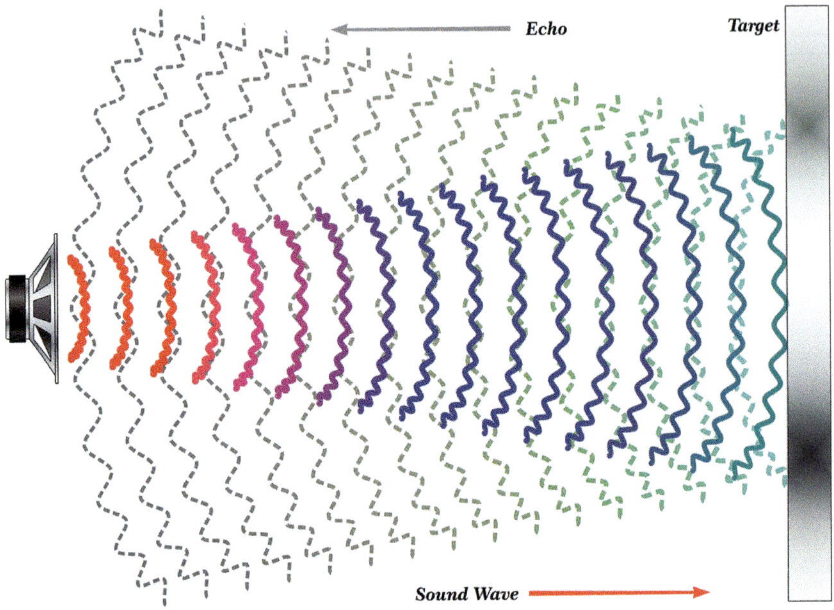

Figure 5.1 Slap echo most commonly appears when sounds reflect off a hard, flat wall.

E. It is common for echo to be hidden or masked by reverberation. It is also common for problematic echoes to become noticeable after a room has been treated—insufficiently or improperly—to reduce excess reverberation.

F. When done properly, it is possible to treat both excess reverberation and echo at the same time, with the same treatment materials. It is all about the physical location of the installed treatment(s).

G. Sometimes, in rooms of all sizes, parallel glass walls or windows create slap or flutter echoes that physically or aesthetically can't be effectively treated. In such cases, the best architectural solution is to tilt or rotate the glass walls or windows enough to minimize the echoes. See more about tilt and rotation in Chapters 6 and 8.

5.4 Reverberation:

A. "Reverberation is sound that persists in an enclosed space, as a result of repeated reflection or scattering, after the sound source has stopped."[1]

B. Reverberation time is typically designated as T_{60}. The letter *T* represents the time in seconds. The subscript *60* represents 60 decibels (dB). T_{60} describes the time it takes a sound to decay or decrease 60 dB in amplitude (volume). A 60 dB volume change is huge, so the takeaways are these:

1. T_{60} represents the time it takes for a moderately loud sound to become almost inaudible. It quantifies the amount of time it takes for the reverberant energy in a room to effectively disappear; that is, until the room is reenergized by a new sound, at which time the process starts over again.

2. While sophisticated hardware and software applications should be used to capture accurate, frequency-specific data, it is common and appropriate to get a rough estimate by using a loud hand clap as the sound source. By counting off the number of seconds it takes for the reverberation to become inaudible, you can get a rough idea of a room's T_{60}. See Chapter 11, Section 11.3, for more on this hand clap test.

   a. It is also likely you will hear slap or flutter echo when doing the hand clap test. When walking around the room, the time and character of the reverberation should remain about the same, but the characteristics of the echoes you hear will likely be different, depending on where you are positioned.

3. You may also see RT_{60} used in place of T_{60}. T_{60} is the correct terminology.

C. Characterizing reverberation time—in the most generic sense—consists of a single unit of time, such as 1.25 seconds. While this single unit descriptor is helpful, it only describes the averaged time decay for 500 hertz (Hz) and 1 kilohertz (kHz).

1. The average of these two frequencies is sometimes called the mid-band T_{60} (T_{mid}) because it better acknowledges the mid-band reverberation time of a room. So, if someone says they have a 1.25 second room, it means the T_{mid} lasts 1.25 seconds. Higher and lower frequencies will usually have shorter or longer decay times.

2. Generally, unless specifically treated otherwise, the larger the room volume, the longer the T_{mid} reverberation time. Therefore, room volume and absorption should be controlled to achieve a T_{mid} that is appropriate and consistent with the room's primary audio activities and applications.

D. Frequency-specific T_{60} guidelines—for each octave center between 63 Hz and at least 2 kHz, and for various audio applications—have largely been ignored in acoustical literature and textbooks. The author's T_{60} Slope Ratio (TSR) thesis[2] provides examples, guidelines, and metrics to help all concerned parties advance and define appropriate reverberation goals. It also offers a simple numeric scoring scale and grading vocabulary from which acoustical design specifications can be initiated and/or evaluated. Examples of how this comes together can be seen in Figures 5.2 and 5.3. This topic is further detailed in Chapter 17.

---

2. https://www.gracenoteds.com/_files/ugd/406ce9_5ceed074e52c4faea76efbacb19e2111.pdf

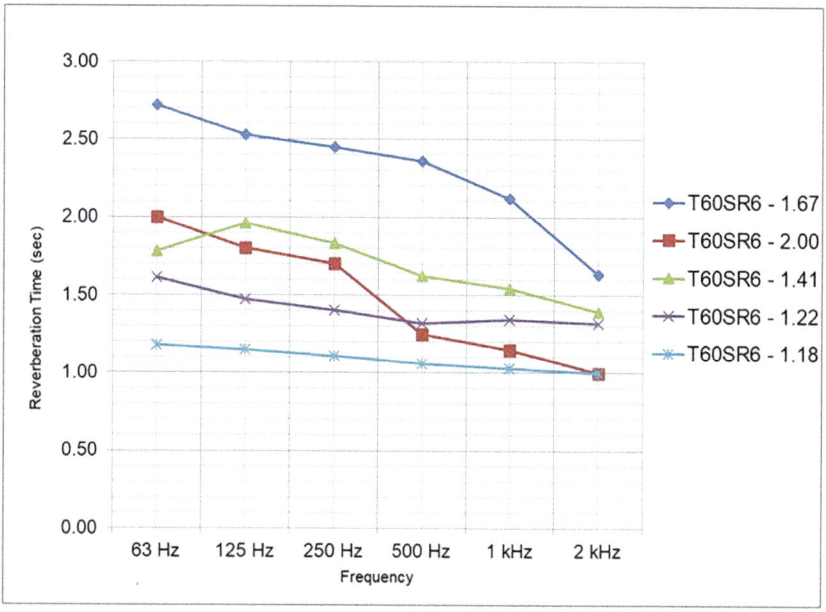

Figure 5.2 This chart shows examples of how widely the reverberation time can vary in different rooms. The TSR thesis suggests that the light blue line at the bottom falls into the optimal range of the sloping T_{60} values because it is the flattest.

	Frequency						$T_{60}SR_6$ Score	Comments	Grade
	63 Hz	125 Hz	250 Hz	500 Hz	1 kHz	2 kHz			
Seconds	2.72	2.53	2.45	2.36	2.12	1.63	**1.67**	Recently Measured RIR	Poor
	2.00	1.80	1.70	1.25	1.15	1.00	**2.00**	From Graph 1 BR Chart	Bad
	1.78	1.96	1.83	1.62	1.54	1.39	**1.41**	Asymmetrical Slope	Fair
	1.61	1.47	1.40	1.32	1.34	1.32	**1.22**	Recently Measured RIR	Good
	1.18	1.15	1.11	1.06	1.03	1.00	**1.18**	Targeted Goal	Optimal

Figure 5.3 This table further details the T_{60} times along with the TSR scores and grades shown on the Figure 5.2 chart. Under the Comments heading: RIR = room impulse response and BR = bass ratio.

E. In most cases, reverberation produces the same diffused sound levels for all listeners. It has no specific direction of energy flow. On the other hand, room modes, which get covered next, produce dramatic level changes from location to location—often separated by only a few feet. Figure 5.4 is a great visual. It tells the story better than a thousand more words could.

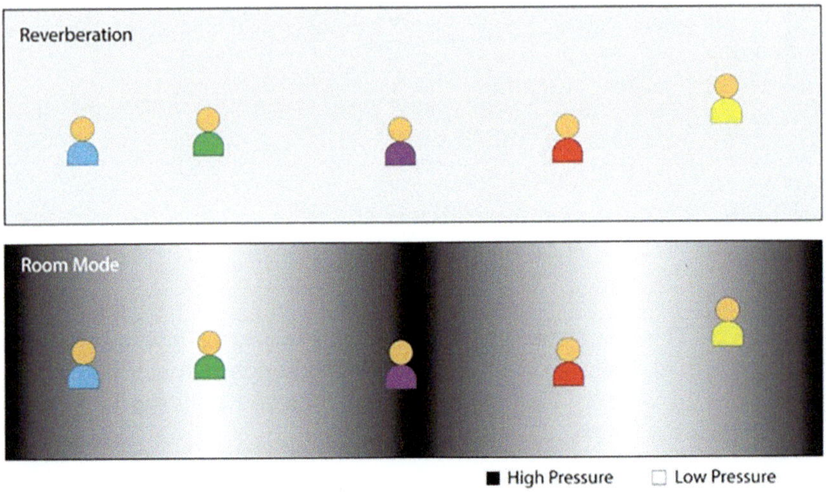

Figure 5.4 Reverberation envelops everyone equally. On the other hand, room modes cause extreme loudness variances in both direct and reflected energy—from one small area to the next. Put in simple terms: high pressure equals very loud. Low pressure equals near silence. *Source*: SynAudCon[3]

F. What is the ideal or optimal reverberation time?

1. There is no ideal T_{60} or T_{mid}. The best we can offer, considering rooms without variable acoustics (VA), are suggestions based on the music genre and the volume (size) of the room. Note: VA is the subject of Chapter 13.

3. Divide and Conquer—The Schroeder Frequency; Pat Brown; https://www.prosoundtraining.com/2021/10/14/divide-and-conquer-the-schroeder-frequency/

a. Leadership in Energy and Environmental Design (LEED) is a global rating system for *green* building structures. LEED guidelines state that the T_{mid} should not exceed 0.80 seconds for lecture halls and 0.60 seconds for small/medium classrooms of less than 10,000 ft^3.

b. For applications that include both spoken words and live, amplified music, the T_{mid} should not exceed about 1.50 seconds. A T_{mid} close to 1.25 seconds is generally optimal for such rooms. Of course, exceptions exist. Venues designed for dedicated applications, such as symphonic or operatic halls, will require a much more detailed analysis and conversation with your favorite acoustical consultant.

c. Example: Figure 5.5 briefly summarizes the results of the research[4,5] done by Niels Werner Adelman-Larson, et al., which concludes that there are two important factors to consider for rhythmic music genres (e.g., rock, pop, jazz, punk, hip-hop, Latin, contemporary worship, etc.): room volume and T_{60} at 125 Hz.

G. Regardless of the room volume—in rooms using amplified sound reinforcement—excess reverberation is the single most likely cause of poor speech intelligibility and/or music clarity.

---

4. https://flexac.com/about-flex-acoustics/#aboutresearchpapers. See the "Research Papers" heading near the bottom of the page for: (a) "Suitable RT in halls for rock & pop music"; (b) "Investigation on acceptable reverberation time at various frequency bands in halls that present amplified music"; and (c) "Importance of bass clarity at amplified music concerts."

5. Rock and Pop Venues-Acoustic and Architectural Design; Niels Werner Adelman-Larsen, et al.; Figure 5.3, pg. 123; Springer

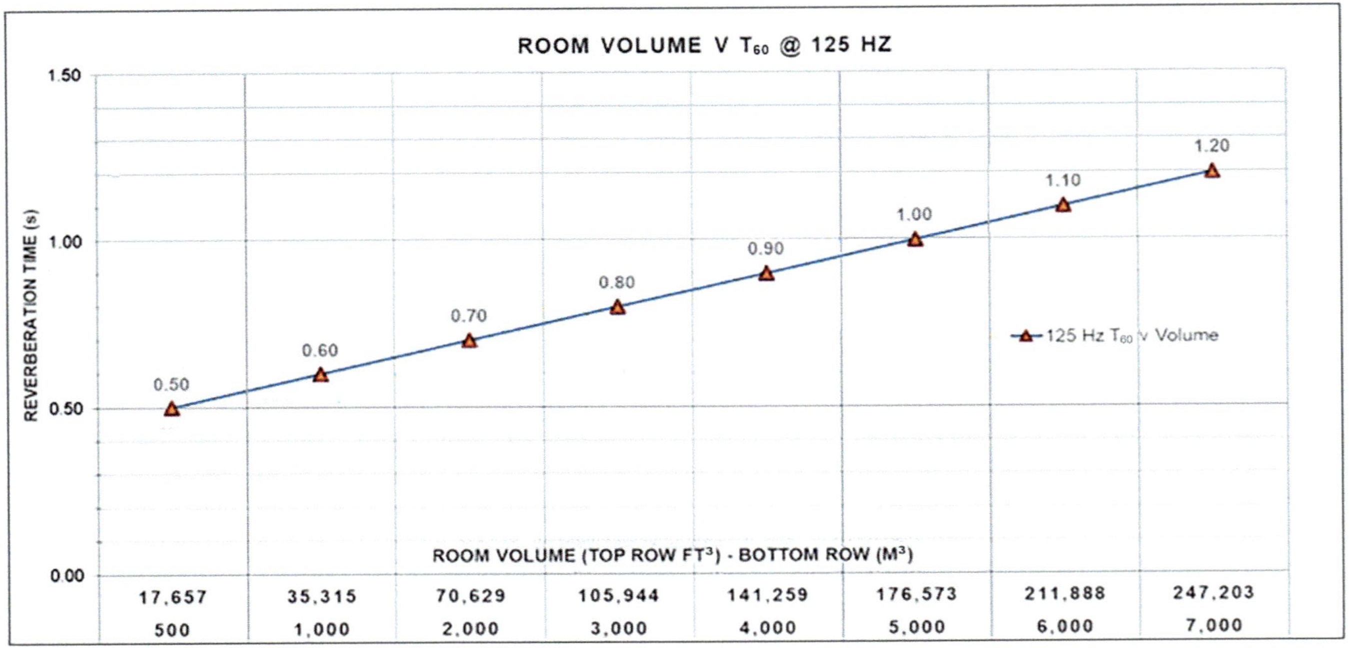

Figure 5.5 This chart shows the trend line that represents suggested 125 Hz T_{60}s for empty halls of various volumes. This trend line continues on a fairly linear path up to 50,000 m^3 and beyond. Example: A 92′ × 55′ × 35′ room has a volume of 5,000 m^3, or roughly 176,573 ft^3.

H. To further research and understand modern reverberation design goals and treatment methodologies, see the T_{60} *Slope Ratio*[2] and *Parametric Acoustics*[6] white papers.

I. Sidebar: You may occasionally hear a recording that sounds like it has a ton of long reverberation, yet the primary signal of the voice or instrument sounds totally clean and clear. This is a mixing trick done in an electronic—not acoustical—environment. The mixing engineer can easily adjust the direct-to-reflection ratio in the mix, so the long reverb time doesn't conflict with the desired intelligibility or clarity of the performance.

   1. To accomplish a similar effect, *coupled rooms or chambers* are designed into concert halls to increase room volume and provide late, longer reverberation.

5.5 Resonance:

A. "Acoustical resonance is a phenomenon in which an acoustical system amplifies sound waves whose frequency matches one of its own natural frequencies of vibration; its resonance frequencies."[7]

B. When applied to architectural acoustics, these resonant frequencies are called room modes, standing waves, or eigentones; all three are synonymous.

   1. Note: For anyone who has been a music student, these are unrelated to musical modes such as Ionic, Dorian, Lydian, or Aeolian, etc.

C. No two rooms are likely to have the exact same collection of room modes, unless they are constructed as exact copies.

D. In rectangular rooms, there are three possible types of modes: axial, tangential, and oblique. Architects need to know very

---

6. https://www.prosoundweb.com/parametric-acoustics-the-case-for-the-parametric-method-of-acoustic-treatment-second-edition/

7. https://en.wikipedia.org/wiki/Acoustic_resonance

little more about room modes. The important exception being that the primary room dimensions should not be multiples of each other. Example: A 48′ × 24′ × 12′ (LWH) room will intensify the most troublesome modal conditions.

1. Axial modes are associated with the distance between any two parallel surfaces (see Figure 5.6). Axial modes tend to dominate and are usually the most problematic.

2. Tangential modes involve four major room surfaces (see Figure 5.7).

3. Oblique modes involve all six major surfaces (see Figure 5.8).

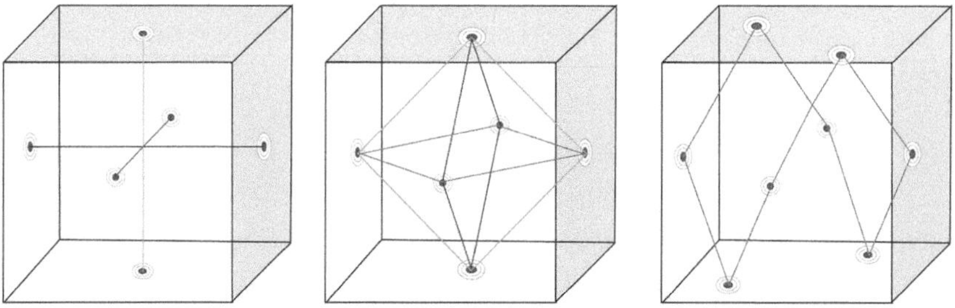

5.6 5.7 5.8

Figures 5.6, 5.7, and 5.8 A schematic representation of the flightpath of axial, tangential, and oblique room modes. *Source*: Atiq Ahmed

4. These modal variations exist in all buildings and they are a potential problem that should be discussed with your favorite acoustical consultant. Let them evaluate and advise.

E. Room modes cause noticeable peaks and nulls (dips) in low frequency (LF) and very low frequency (VLF) response. When two or more waves meet and are in phase with each other at a specific frequency, there will be a peak (increase in level or volume) in response. When they meet and are out of phase with each other, they cancel each other out causing a dip or null, which is a decrease in level or loudness (see Figure 5.9).

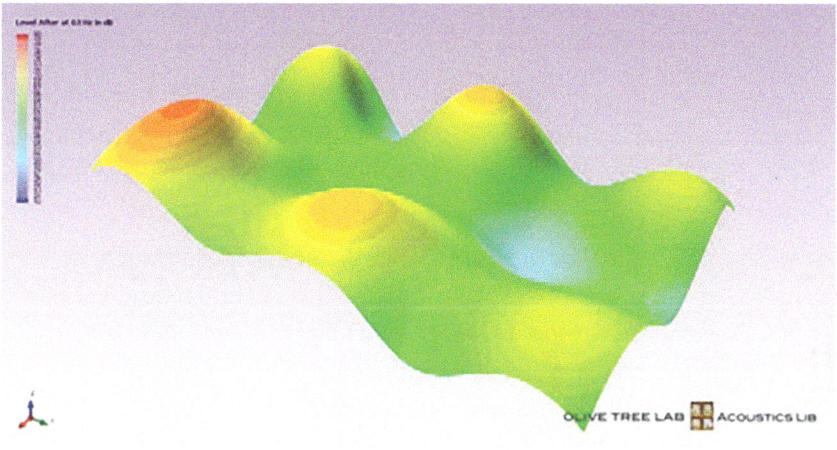

Figure 5.9 This 3-D graphic shows an example of the huge range of energy variance caused by unmanaged room modes. The amplitude range between the orange peaks and the light blue valleys is nearly 30 dB. For reference, a 10 dB change in loudness is generally considered to be twice or half as loud. *Source*: Olive Tree Labs Suite at mediterraneanacoustics.com

F. Medium and large rooms:

1. VLF and LF room modes force certain musical notes to sound much louder or softer at various locations within a room.

2. Dominant modes (primarily the axial type) are problematic because they produce dramatic level changes from area to area—often separated by only a few feet. This helps explain why one person will complain that the bass guitar is overwhelmingly loud, while someone sitting just a few feet away may observe that it is not loud enough.

3. To add complexity to all this, modes are shape shifters. If the music changes to a different key, the notes that are too loud or soft are likely to change positions too. It is not that the modes change position but that the peaks and dips develop in different locations for each modal frequency.

   a. In plan, the area affected by these extreme loudness changes can range from about 25 to 100 square feet. In elevation, the range may be much less, depending on the ceiling height.

4. Note for medium and large rooms: Even though room modes are not technically reverberation, they may resonate longer than all other low-, mid-, and high-frequency reverberation. This means they still can *muddy up* a musical presentation and should be properly treated in order to improve the *music clarity* of a sound-reinforced performance (see Figure 5.10).

G. In small rooms, these same modal anomalies reach into the low- and mid-frequencies. The smaller the room, the higher the frequency range of potentially problematic modes. Because of this, small rooms—such as conference or huddle rooms—can cause uneven sound reception in the fundamental range (80 Hz–300 Hz) of both male and female voices.

1. The takeaway: Given any room's primary dimensions, certain fundamental voice frequencies may be much louder or much softer depending on where the listeners are seated or where microphones are placed. The spacing of these extreme variances can be as little as a few feet.

2. Note: Small rooms rarely have true reverberation. What might be perceived as reverberation is usually high-density modal resonance and/or flutter echoes.

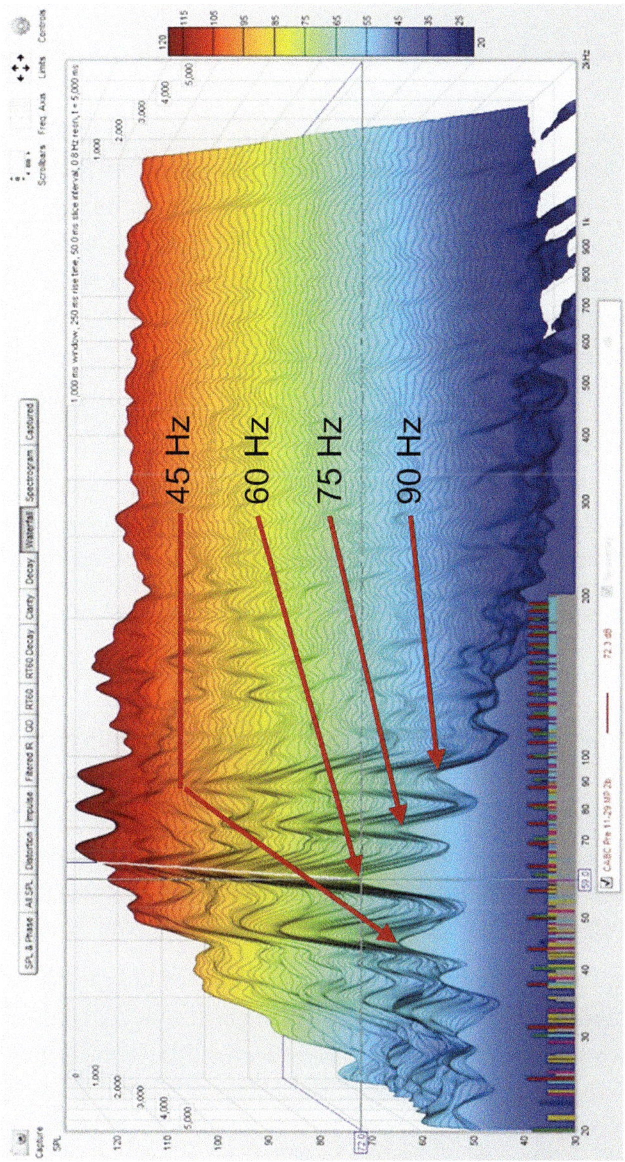

Figure 5.10 This is a waterfall plot showing dominant modal behavior at 45 Hz, 60 Hz, 75 Hz, and 90 Hz. Note that the resonant modal behavior is stronger and lasts longer than the decaying reverberant energy at all the surrounding frequencies. This measurement was taken in a 400,000 ft^3 room. When viewing this 3-D plot, time increases back to front; the frequency range is 20 Hz to 2 kHz; and the amplitude ranges from 128 dB sound pressure level (SPL) (in dark red) to less than 40 dB SPL (dark blue) above 500 Hz. Graphic generated using Room EQ Wizard[8] software application.

8. https://www.roomeqwizard.com/

H. "Dealing with acoustic modes is accomplished by applying frequency-specific absorption to at least one of the boundaries, thereby minimizing the reflections off of it, so there is less energy to combine or cancel. While corners are not a complete solution, they do offer the advantage of being at the end of two, or even three, of the room dimensions, so there is a lot of benefit in treating those areas. Sometimes there are modal issues that require acoustical treatment of the walls or even the ceiling over your head. With these scenarios, treating corners alone would not be sufficient."[9]

I. For anyone wishing to take a slightly deeper dive into the behavior of room modes and how they are formed, take a look at this University of Southampton blog.[10] Their animation provides a good illustration of how modes (at two different frequencies) are developed and move around in a room.

5.6 Perhaps a few metaphors might help everyone get a better feel for the behavioral differences between echo, reverberation, and resonance.

A. Visualizing echo: Echo, be it flutter or slap, is a bit like walking into a room with two parallel walls that are largely covered by mirrors. If you can focus on one, while tuning out all the others, you see a discrete reflection of your body. Because the photons are traveling at the speed of light, your reflection represents a singular, instantaneous slap echo of your image.

1. Now slow those photons down to the speed of sound and delay the image so it is no longer in sync with your movements; any activity will be seen a little later in time than the actual movement. Imagine a video with poor lip sync quality when someone is talking. This comes close to describing the effects of slap echo. It can be a major distraction to all.

---

9. https://www.gikacoustics.com/what-are-room-modes/

10. http://blog.soton.ac.uk/soundwaves/standing-waves/5-room-modes/

2. Next, stand back and look around. You'll begin to see a nearly infinite number of copies of your image, a visual feedback loop (see Figure 5.11). Flutter echo is made of repetitive, reflected sound waves that slowly decay over time. And they don't stop immediately when the sound source stops.

Figure 5.11 This visual feedback loop is somewhat like what flutter echo would look like if we could see it. Each repetition arrives a little later in time and with a little less energy. The net result is an image or sound that is hard to decipher or understand.

3. The visual feedback loop is made of photon rays that don't decay over time, but simply stop when you leave the room or eliminate all visible light. Flutter echo tends to smear or distort the quality of the sounds you want or need to hear clearly. Again, a major distraction to all.

5.7 Visualizing reverberation: In the mind's eye, reverberation can be made visual if we convert it to fog (see Figure 5.4, top). In a room with optimal, frequency specific reverberation characteristics, there would be very little or no fog. As the reverberation time becomes longer and longer, the room fills with more and more fog, until nothing in the room is visible beyond a few feet. For this metaphor, substitute the lack of visibility for lack of intelligibility and/or music clarity.

5.8 Feeling modal resonance (see Figure 5.4, bottom): Imagine sitting in a large room with an HVAC system that is dumping both warm and cold air into the room at the same time. Also imagine that the ducting is laid out such that every 10 feet or so, there is an overhead vent supplying either very warm or very cold air. In between the vents, you may feel comfortable, but if you're sitting close to, or directly under one of the vents, you'll probably be too warm (i.e., too loud) or too cold (i.e., barely inaudible). Here we substitute warm and cold air for the wild swings in volume caused by a room's modal behavior.

A. But wait, there's more. Let's not forget the shape-shifting factor and its Goldilocks principle: Now the system has a random number generator programmed into each duct and every few minutes the ducts switch to something different; too hot or too cold, but never just right. If they actually existed, no one would put up with these shape-shifting thermal conditions for very long. Yet, we are often forced to live with these very real acoustical irregularities.

5.9 Unmanaged echo, reverberation, and/or resonances are often our greatest challenges. They represent the net acoustical outcome of unknowing or uncaring owner and/or architectural decisions, or the misalignment of a venue's design and primary application(s).

CHAPTER 6

THE THREE ACOUSTICAL TOOLS

6.1 There are three primary acoustical *tools*, or methodologies, used to manage echo, reverberation, and resonance. Those tools are: reflection, absorption, and diffusion (see Figure 6.1). These are not actual tools, but rather they are adnouns that represent the acoustical properties and interactive behavior of various finish materials. These three words translate into product categories, materials, and application techniques that can be used to regulate how sound energy behaves in an enclosed space.

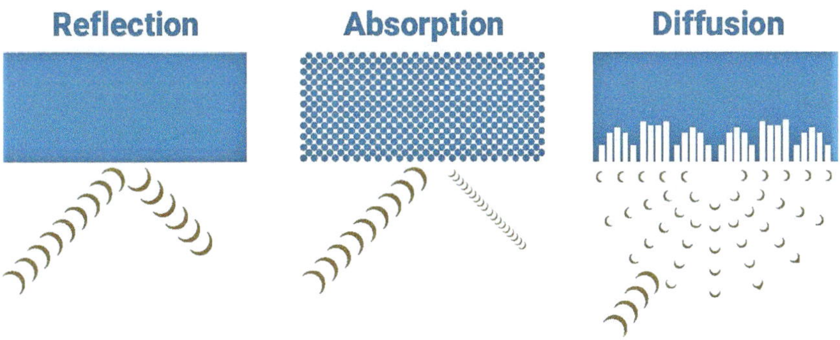

Figure 6.1 Simplified views of the residual effects of reflection, absorption, and diffusion. Note the size of the incoming sound waves at the lower left of each panel and the size and behavior of the sound waves after contacting the various materials.

6.2 Reflection: With exceptions (see Figure 6.2), reflective surfaces are typically not specified as mitigating acoustical treatments; rather, they tend to support and sustain sound energy longer than may

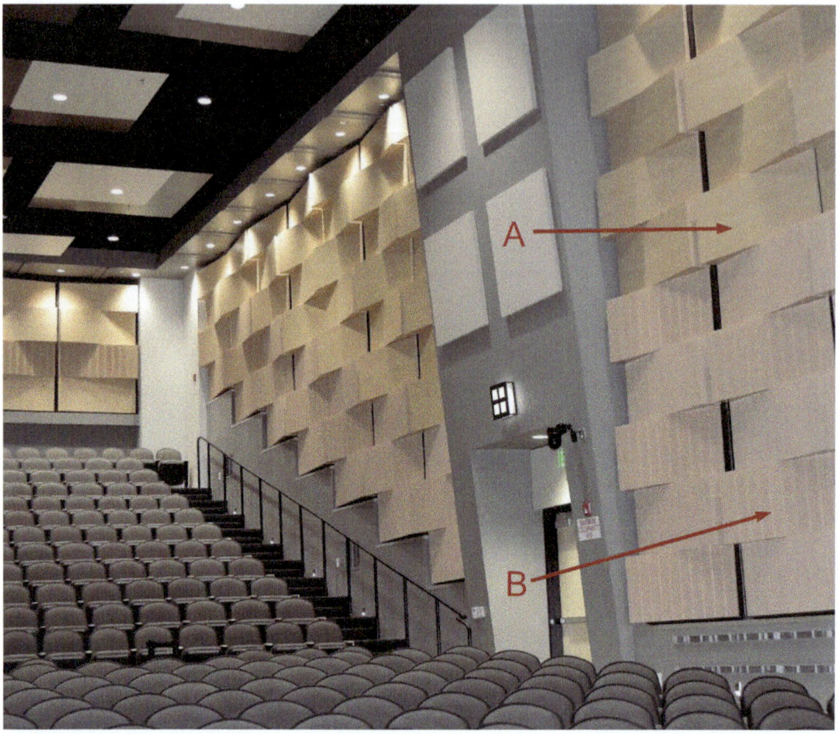

Figure 6.2 The splayed, reflective [A] and perforated [B] panels on the side and rear walls provide unique aesthetic and acoustical performance. It may be difficult to see in this photo, but such panels visually present as one finish but there are both reflective and absorptive modules. Absorption is achieved via the perforations with absorptive filler behind. *Source*: RealAcoustix.com

be necessary or desired. Many very common structural and finish materials are reflective, so they should be understood in the acoustical context of an overall room design.

A. Reflective materials are generally smooth and rigid. Common examples include concrete, tile, glass, wood, brick, metal, and sheetrock.

1. If you shine a flashlight at a mirror, you get a reflection, or bounce, which is nearly identical to the original light source. This is similar to what happens when sound bounces off smooth, flat, acoustically reflective surfaces.
2. When sound energy strikes a flat reflective surface, the sound bounces off with nearly the same energy it had before striking the surface. These are sometimes described as *specular* reflections.

B. The application of reflective finishes ranges from good to bad, based on shape, size, and location. Acoustically, here's how to prioritize specifying reflective finish materials:

1. The good:
    a. Almost anything with a convex shape, regardless of size or location. Why? Because convex reflectors are essentially sound scattering devices and, acoustically, sound scattering is almost always a good thing.
    b. Any other complex geometry that redirects sound energy away from its source and away from other parallel surfaces that are also reflective. This is the principle behind the reflective panels shown in Figure 6.2.
    c. In venues built mainly for the presentation of acoustic music, reflection is often used as a tool of amplification via finish materials and shapes that sustain sound energy and promote useful early reflections and reverberation (see Figure 6.3). Such reflective surfaces support the weak sounds that are associated with choirs and acoustic instruments.

Figure 6.3 This photo shows a Wenger, Maestro Shell 5, reflective stage shell. Note that from a distance, this presents as a large concave reflector, but upon closer inspection, each segment has a slight convex curvature. This design takes simultaneous advantage of both concave and convex topologies. *Source*: Wenger Corporation

2. The acceptable but potentially problematic if unmanaged:

   a. Almost any hard surface that is flat and parallel with another flat, reflective surface of meaningful size, that is, greater than about 10 ft^2. These conditions promote flutter echo. To minimize flutter echo, reflective parallel surfaces can be partially or fully covered with absorptive and/or diffusive/scattering treatments. Refer to Chapter 5, Section 5.3, for more on flutter echo.

b. Alternatively, those large surfaces can be tilted, rotated, or staggered by at least a 1:12 offset (~5.5°). A 2:12 offset (~11°) is preferred. Here, the treatment is not added products or materials, just simply the slight movement or repositioning of flat reflective structures.

c. For those who aren't architects or general contractors, here is what these ratios mean: For every 12″ of linear distance, a 1″ to 2″ offset must be applied. In other words, a wall that's twelve feet wide needs to tilt or stagger by one to two feet.

d. When opposing, mirror image walls are offset, the offset effectively doubles the stagger, rotation, or tilt. More is usually better.

3. The bad:

    a. Almost any concave, reflective shape. Depending on location and radius, reflective concave shapes can be extremely problematic. They collect and refocus sound energy, generally amplifying and concentrating sound in unfavorable ways.

    b. However, small concave finishes that are located high enough above the floor will have focal points well above the listeners, thereby providing useful scattering.

    c. If not carefully located, analyzed, and treated, any form of a curved/faceted wall—and/or arched or domed ceiling—is likely to introduce significant acoustical anomalies, such as those listed in the following sections d, e, and f.

d. For listeners, the refocused sound energy causes audible *hotspots*, depending on where one is seated—under or within the focal radii of the curve, ellipse, or dome. And, depending on seating location, the refocused energy may also be heard as an audible slap echo.

e. If any microphones are positioned within the focal radii, audio feedback squeals can easily become a common occurrence and very difficult to mitigate electronically.

f. Hard flooring and/or seats compound the problems because of the recirculating, specular reflections that occur.

C. To be fully effective at reflecting low frequencies (LFs) or very low frequencies (VLFs), finish materials must be very dense and rigidly mounted. Common construction methods are generally insufficient.

1. Think about designs that use concrete or block walls or that require double or triple layers of sheetrock with stud spacing no greater than 12″ on center. There are only a few audio applications that require and benefit from such density. Symphonic concert halls that only present unamplified, acoustic music are one of the rare exceptions.

D. With all that said, hard reflective finishes are common, expected, and generally unavoidable. Therefore, the architect's challenge is to develop designs that recognize the strengths and weaknesses of reflective finishes and, if necessary, acoustically treat them to optimize their performance within the overall desired or required acoustical signature of a venue.

6.3 Absorption: Sound energy is absorbed when various materials and structures provide a means of converting its kinetic energy into heat. This happens when the energy strikes an absorptive or resonant wall, floor, ceiling, person, panel, boundary, or barrier of some type. The result is that some fraction or percentage of the sound—at each discrete frequency—will be converted to heat, thereby reducing its audible energy.

A. Before we dig in further, it's helpful to briefly touch on the physics behind how absorption tools interact with sound. I realize these next several paragraphs may seem too deep for what's supposed to be a nontechnical commentary—bear with me. The rationale is to also provide the *why*. Acoustics are already complicated and perplexing enough. Without touching on why sound energy is such a challenge to manage, there's little hope for advancing the subject into meaningful, proactive action in the dominions of architecture, interior design, and ownership.

B. Due to the enormous range of frequencies and corresponding wavelengths we can hear, sound energy is captured and dissipated in two distinctly different ways. One is *velocity-based* (VB) absorption, the other is *pressure-based* (PB) absorption. Because of this, two entirely different methodologies are required, depending on what range of frequencies need treatment.

   1. To avoid any confusion, we are talking about absorption that is attained as a result of slowing down the *particle velocity*—at the micro level—within the wave energy of an oscillation, not a change in the speed of sound.

C. It is all about the Schroeder frequency.

   1. Sound behaves differently depending on frequency and the environment in which it is propagated. Dr. Manfred Schroeder[1] referred to the frequency at which rooms go from being resonators to being reflectors/diffusors

1. https://en.wikipedia.org/wiki/Manfred_R._Schroeder

as the *crossover frequency*. We now call it the Schroeder frequency, with F_S as the designator.[2]

2. There is a simple formula for determining the Schroeder frequency: $F_S = K\sqrt{T_{60}/V}$. Where the reverberation time (T_{60}) is the mid-band (T_{mid}) reverb time, V is the room volume in cubic feet or cubic meters, and K is a constant. $K = 11{,}885$ when using U.S. Customary Units (USC), or 2,000 when using the International System of Units (SI). The environment being referred to is an enclosed room, regardless of size. Together, room volume and its T_{mid} reverberation time determine the resonant frequency (F_S).

3. F_S defines the initial (highest) frequency, below which all lower sounds behave like waves. Once you know F_S, it is then necessary to multiply that frequency by four to determine the upper limit ($4F_S$) of the *transition zone.* $4F_S$ establishes the frequency at which all higher-frequency sounds behave like light rays.

    a. The transition zone consists of frequencies with ambiguous behavior—neither fully resonant/modal, nor completely diffuse reverberation. This is not a hard transition but a gradual one, which occupies about two octaves of sound and includes all frequencies between F_S and $4F_S$.

4. Examples of how room volume influences F_S:

    a. A small 1,000 ft^3 conference room, with dimensions of $10' \times 12.5' \times 8'$ (LWH), and a T_{mid} of 1.0 second, will exhibit resonant behavior (F_S) at 376 hertz (Hz) and below. Translation: the fundamental frequencies of most male and female voices

2. Architectural Acoustics, 2nd Edition. Marshall Long. p. 327. Elsevier Academic Press.

will easily fall into the frequency spectrum where modal anomalies may dominate audio propagation or reception. For a refresher on this topic, see Chapter 5, Section 5.5.E.

b. Continuing with the same T_{mid}, a small 4,000 ft^3 classroom will have an F_S of 188 Hz. At this frequency and below, the sound of many adult male voices will be compromised by the room's modal behavior.

c. A small music rehearsal room might be 8,000 ft^3 and have a T_{mid} of 1.50 seconds. This will result in an F_S of 163 Hz.

d. A medium-sized church, with a volume of 100,000 ft^3, and a 1.25 second T_{mid}, will result in an F_S of 46 Hz.

D. Generally, mid frequency (MF) and high frequency (HF) sound energy with *ray* behavior above $4F_S$ is most efficiently absorbed with porous, VB treatment materials like those listed in Section 6.3.F.3 of this chapter. LF and VLF sounds with *wave* behavior at and below F_S are most efficiently captured using nonporous PB membrane treatments, which are further defined in Section 6.3.H.3 of this chapter. These are important distinctions that explain some of the complexities and many of the misunderstandings related to the acoustical treatment process. I call this the wave/ray duality of sound.

E. As sound energy moves through a room, the particle velocity within each wave is greatest as the air molecules are in transition between the peaks and valleys of each sinusoidal cycle (see Figure 6.4).

1. For VB absorption, "Most of the absorption comes from the way the sound waves squeeze through the labyrinth of tiny, interconnected pores within the material. As the sound moves through the pores, it causes

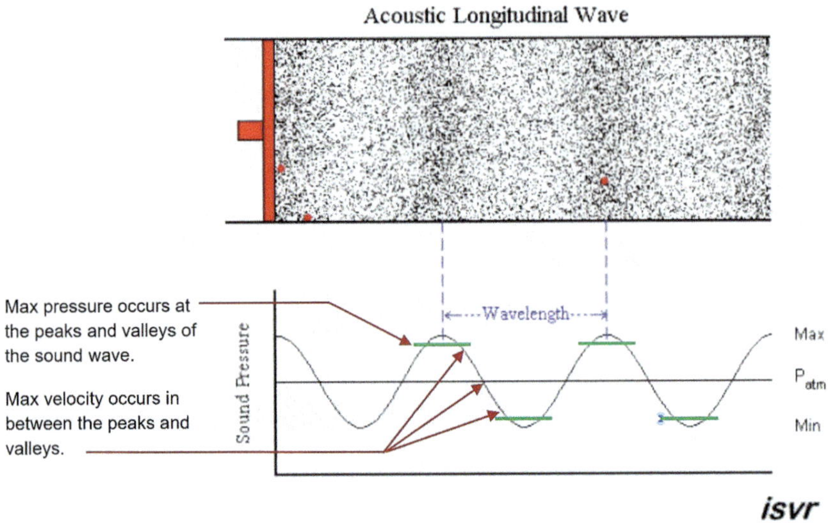

Figure 6.4 This graphic shows approximately where sound waves reach maximum pressure and velocity. *Source*: Teachoo.com

air molecules to drag along the interior surfaces of the material. It takes effort to drag the air molecules across the interior surfaces. This effort turns the vibrations of the molecules into heat, which are then dissipated in the material."[3]

2. VB absorption treatments are most effective when the front face of the material is mounted one-quarter wavelength or greater from a rigid boundary. Example: One-quarter wavelength of 1,000 Hz is roughly 3″.
   a. This recommendation doesn't mean that a 3″ absorber is required to absorb 1,000 Hz sounds; it means that if you need to absorb a certain amount of energy at that frequency, mounting the front face 3″ off the wall will maximize the efficiency and effectiveness of the specified treatment.

3. Choosing & Using Porous Absorbers, Trevor Cox. Sound on Sound Magazine. July 2015. https://www.soundonsound.com/reviews/choosing-using-porous-absorbers

b. Also, at 1,000 Hz, the quarter wavelength spacing can be accomplished with a 3″ panel mounted directly to the wall; a 2″ panel mounted on a pair of 1″ fir strips to create a 1″ air gap behind; or even a 1″ panel, mounted on 2″ fir strips. In this example, all higher frequencies (shorter wavelengths) benefit from the optimized efficiencies of the spacing off the wall. Only lower frequencies will gradually provide less efficiency and performance.

3. Once again, these details aren't anything architects need to master; just some important concepts to be aware of.

F. VB absorption:

1. Soft, porous materials have VB absorptive properties and do not make good PB treatments. Because of this wave/ray duality, materials, products, and techniques that effectively absorb sound energy above about 250 Hz are generally least effective at absorbing energy below 250 Hz. The inverse relationship is also true.

2. Regardless of size, type, cost, location, or appearance, all treatment materials are only partially effective at absorbing sound. Each type and thickness have different absorption coefficient contours (see Chapter 4, Section 4.4). This is a fancy way of saying each material absorbs either more or less sound energy across the various frequencies of the audio spectrum.

3. VB absorption examples:

a. Mineral wool, rock wool, spun glass insulation, and medium- and high-density fiberglass panels are very useful VB absorption tools when properly specified, located, and installed.

b. Exposed, R-11, R-19, and R-30 batt insulations are effective tools too. However, these materials deliver little or no absorption if fully enclosed in a wall or placed above a hard-lid ceiling. When specified as a sound absorber, these thermal insulation materials must be exposed to the air and sound energy within the room. If necessary for aesthetic reasons, finish with an acoustically transparent fabric or perforated/slat wall paneling.

c. Indirect exposure is useful when open vaults, cavities, soffits, and traps are created in a room. In these areas, batt insulation can be used to fill cavities or to line walls, as necessary. Additional finish materials are generally not required in areas that are invisible to the public.

d. If faced, the soft, fuzzy side of the insulation should always be mounted outward, so it is exposed to the air as much as possible, and not facing the structure to which it is attached.

4. Not all *insulation* works as a sound absorber.

   a. In the construction trade, the term *insulation* describes two structurally different types of material—one is useful to the acoustician, and one is not.

   b. Closed cell, extruded polystyrene (XPS) rigid foam insulation should never be specified or used as a sound absorber. This rigid foam absorbs little or no sound.

5. Semi-rigid fiberglass panels are very effective VB sound absorbers and are one of the most commonly specified types.

   a. These products are typically specified with a density of either 3 pcf (pounds per cubic foot) or 6 to 7 pcf. Other densities exist but are less commonly required.

b. Fabric-wrapped wall panels, sometimes referred to as *sound soak* panels, are usually constructed from 6 to 7 pcf fiberglass or rock wool boards (see Figure 6.5). While maintaining similar MF/HF performance, the thicker the panel, the lower the frequency range of effective absorption. However, LF/VLF performance limitations, aesthetics, and cost often limit the practicality of such panels if they are more than 3″ thick.

c. Such panels, commonly sold in 1″ and 2″ thicknesses, are typically attached directly to a wall or, preferably, furred off the surface by 1″ to 3″. The furring helps move the front face closer to the max particle-velocity zone at slightly lower frequencies.

Figure 6.5 Fabric wrapped fiberglass panels are available in countless sizes, shapes, and colors. This graphic shows one creative option. The offset panel thicknesses are intentional. *Source*: Lamvin.com

d. When you suspend such panels from the ceiling as hanging baffles or clouds (see Figure 6.6), they provide extra performance and value because both sides are exposed to the sound energy that passes by. Installing such panels as baffles or clouds effectively doubles the square footage of the exposed treatment. A 4′ × 8′ baffle potentially represents 64 ft^2 of VB absorptive treatment.

Figure 6.6 The round blue, green, and white panels are 6 to 7 pcf absorption clouds. *Source*: Lamvin.com

e. Note: Baffles and clouds are essentially the same thing—baffles hang vertically, while clouds hang horizontally.

6. Duct liner is an example of a 1.5 to 3 pcf absorption material. It provides useful VB absorption when applied externally, or anywhere in a room that is directly or indirectly exposed to sound energy. However, when installed inside an air duct, it has little or no impact on the sound energy that strikes the outside of an HVAC duct.

7. Be sure to check the fire rating of any/all VB absorption materials you might consider using. Most should have a Class A or Class 1 rating. If you can't find a lab-tested fire rating report, do not allow it to be installed.

8. Perforated (perf) panels (see Figure 6.7) are another excellent choice to consider when some amount of absorption is required in areas where hard, smooth, flat and/or concaved surfaces are to be specified.

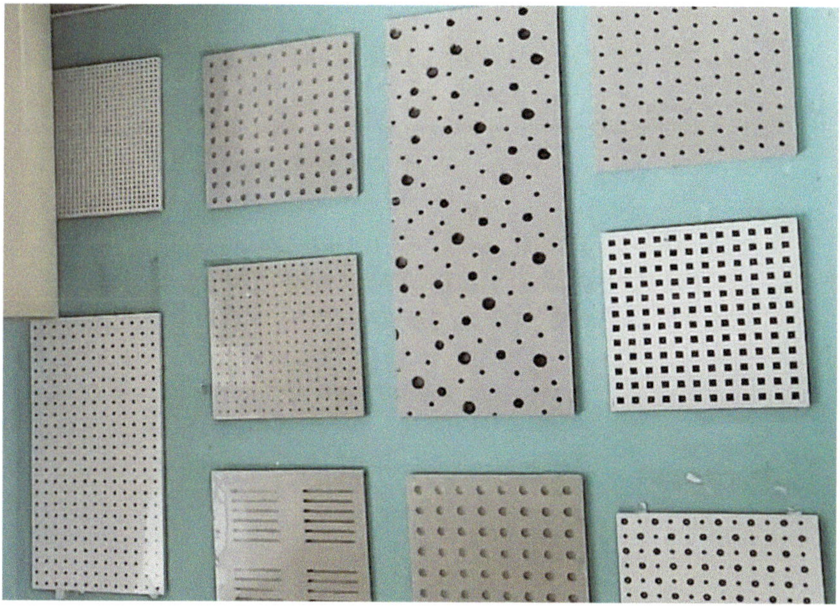

Figure 6.7 Samples of various perforated gypsum board patterns. Many more are available.

a. Perf panels offer a very effective combination of architectural, aesthetic, and acoustical properties that are hard to ignore. They also offer a blending of all three acoustical tools: very good VB absorption; a modest amount of reflection, and possibly a little HF scattering.

b. To qualify as a VB acoustical surface, there is one detail to keep in mind: Perf panels must always be mounted with airspace behind them, and that airspace must be lined or filled with unfaced mineral wool, spun glass insulation, or similar absorptive material.

c. Airspace guidelines for medium and large venues: 2″ = bare minimum, 4″ = nominal, 6″ or more extends the frequencies of LF absorption a little. For small rooms, the bare minimum is 1″.

d. Generically, perf panels have a specification labeled *percent of open area*. This number indicates the percentage of the surface that is exposed to the absorptive material behind the panel. The higher the percent of open area, the greater the potential absorption. The typical range for this application is 20 to 40 percent open area. Iperf.org offers even more detail with its transparency index (TI). This index incorporates the number of perforations per square inch, perforation diameter, sheet thickness, shortest distance between holes, on-center hole spacing, percent open area, and sound frequency. The typical range for the TI application is 35 to 65 percent open area.

e. Regardless of size or shape, the perforations effectively convert a reflective surface into one that is reasonably absorptive.

f. In addition to gypsum board, perf panels are available in a variety of other material compositions, including metal (see Figures 6.8 and 6.15), polypropylene, and fiberboard. The McNichols[4] company is a good resource for perf metal and fiberglass sheeting.

Figure 6.8 Gymnasiums present many acoustical challenges. At St. Augustine High School in San Diego, the architect (Domusstudio Architecture) was willing to change this metal sheeting material—above the bleachers on each side—from fully reflective to thousands of sq. ft. of perforated metal. Six inches of unfaced, R-19 batt insulation was installed above the perf sheeting. Along with other absorptive treatments behind the bleachers and baffles hanging from the ceiling, this turned out to be one of the best-sounding multipurpose gyms in the region.

g. There are a wide variety of hole and slot sizes, shapes, and patterns to choose from. Most panel types can be painted or wrapped with acoustically transparent fabric, as needed, to best tie in with a room's color and finish schedules.

4. https://www.mcnichols.com/perforated-metal

9. In most cases, a perforated metal roof deck is a much better solution than a solid metal roof deck. A perf roof deck provides a very large, well-distributed absorption option that will not likely appear as a line item in the acoustical budget.
    a. Be sure to specify a VB-type sound-absorbing filler material, not the XPS-type rigid foam (thermal only) filler.
10. Micro-perf panels are what the name suggests: panels with thousands of very small slits or pin holes. Figure 6.9 shows an example of a micro-perf application that

Figure 6.9 Until recently, glass has been acoustically problematic. If your client insists on a meeting room with a large glass wall, this image shows what's currently possible using clear plastic, micro-perf absorption sheeting. *Source*: RealAcoustix.com

solves an age-old acoustical problem—installing absorptive treatment in front of a glass window or wall.

a. Micro-perf panels come in a variety of materials and finishes.

b. Take care not to fill any of the holes when painting opaque micro-perf panels.

11. Slat absorbers (sometimes called diffraction slot absorbers) offer many of the advantages of perf panels with the added benefit of delivering greater MF and HF scattering (see Figure 6.10).

Figure 6.10 An example of the visible part of a slat absorber. This type of panel assembly can provide both absorption and scattering. No sound energy is absorbed by the wood slats; rather it is absorbed by the VB filler materials that must be installed behind the slats. *Source*: RealAcoustix.com

a. Slats can be oriented horizontally, vertically, or diagonally.

b. The width and depth of the slats and gaps determine the frequency spectrum that will receive maximum diffusion/scattering. See more on diffusion and scattering in Section 6.4.

c. Like all perf panels, slat absorbers require that a VB absorptive material be installed behind the slats. The slats themselves only provide finish aesthetics and scattering.

12. Slat absorbers can also be configured to absorb LFs and VLFs. To achieve this added performance, the slats must be installed as the front face of an otherwise sealed box consisting of rigid sides and back with some form of VB absorption inside. Further, the depth of the sealed box or cavity and the size and spacing of the slats will determine the LF performance. See Chapter 13, Section 13.5.C, for a modern example of this application.

G. There are many other commercially available VB absorption products. They include open-cell foam, spray-on cellulose, recycled cotton, bonded wood shavings, polyethylene terephthalate felt, carpet, open-cell metal foam (see Figure 6.11), acoustic plaster, and probably a few more that are not coming immediately to mind.

H. PB membrane/resonant absorption:

1. LF and VLF energy that is exhibiting wave behavior at F_S and below is most efficiently absorbed using PB techniques and devices.

2. There are only a few LF/VLF absorption techniques that are definable and reasonably predictable. The two most widely accepted methods are classified as *diaphragmatic* membrane absorbers and Helmholtz resonators.

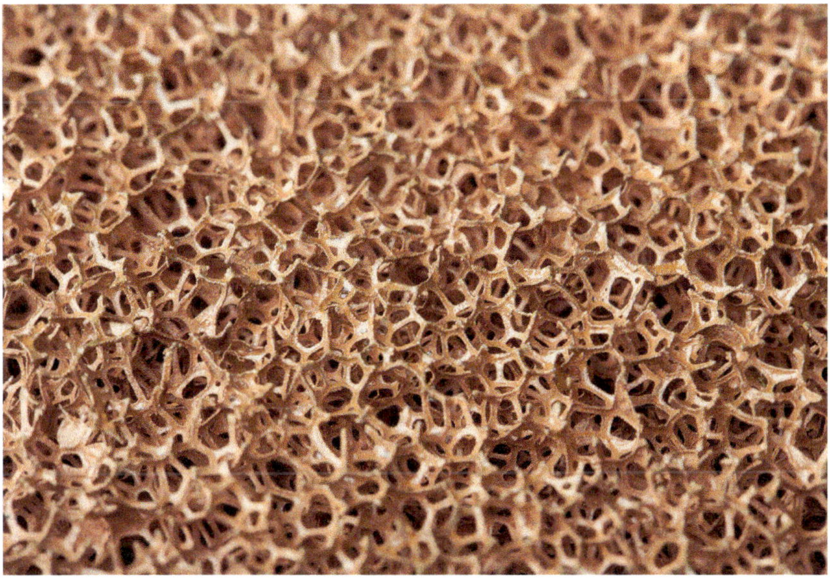

Figure 6.11 This photo shows the structure of open-cell metal foam, as seen through an electron microscope. *Source*: Shutterstock

3. Diaphragmatic absorbers are *limp mass* structures that absorb a moderately narrow frequency range of reverberation and/or modal energy. They can be designed to perform within a fairly specific range of frequencies, as necessary. They have the following properties:
    a. A rigid, sealed, rectangular box or cavity.
    b. Well-defined dimensions and materials.
    c. Typical sizes ranging from 2′ × 2′ × 6″ (hwd), up to 4′ × 8′ × 12″.
    d. The front face membrane of the enclosure—the *working surface*—consists of a thin, unperforated panel of wood, plastic, metal, or other flexible material.

   e. Though not technically accurate, these PB topologies are often euphemistically called *bass traps*. However, not all bass traps are designed and built following diaphragmatic or Helmholtz topologies. Buyer beware!

4. When sound energy moves through a room, its molecular pressure will always be greatest when its long wavelengths reach a rigid boundary. At that point, the velocity is essentially zero because once reaching the boundary, the air molecules at the front of the wave build up and compress as they briefly stop moving in their original direction. Any reflected energy is forced to change direction.

5. Diaphragmatic panels are considered the most effective means of PB absorption (see Figure 6.12). Such devices are not for the DIY crowd. For proper *tuning*, they require very specific sizes, materials, construction methods, and tolerances. Each design modification also requires testing in a very large reverberation chamber to verify performance.

6. Diaphragmatic PB treatments are most efficient when their hardware enclosures are recessed—or even partially recessed—so the front face is flush with its surrounding rigid boundary, or when the back panel of the enclosure is mounted directly against a rigid boundary such as a floor, wall, or ceiling. Why? Because when properly located, this puts the *working membrane* within a maximum pressure zone in the room.

7. When properly constructed and deployed, the vibration of the working surface converts the LF/VLF energy into heat via the friction that exists between the panel and its mounting attachments, the energy loss that occurs through the mechanical strain and deformation of the

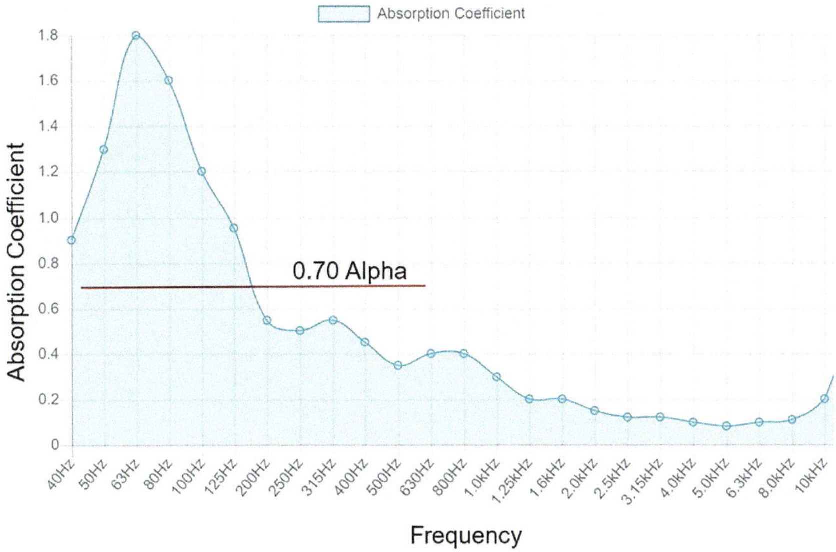

Figure 6.12 This graph shows the alpha performance of a lab-tested diaphragmatic absorption panel. It is the RealAcoustix–BassMod 3248. Any alpha above 0.70 is considered a useful value. This product appears to offer nearly 2.5 octaves of effective treatment between 40 Hz and 125 Hz. At 63 Hz, this 32″ × 48″ panel can provide 1.8 sabins of absorption per ft^2. And yes, recent lab research has shown that it is possible to have an alpha above 1.0. *Source*: RealAcoustix.com

flexible panel, the modulation of the internal air spring, and minimization, which is the small amount of energy cancellation that is caused by reradiation interferences.

a. Note: Gypsum wallboard, glass, wood, and metal can also be resonant absorbers. While these materials will provide some amount of PB absorption, their low alphas make them inefficient treatment tools when amplified music is performed or reproduced (see Figures 6.13 and 6.14). Such generic construction materials contribute to the overall *room absorption* but are not generally considered to be primary or remedial acoustical tools.

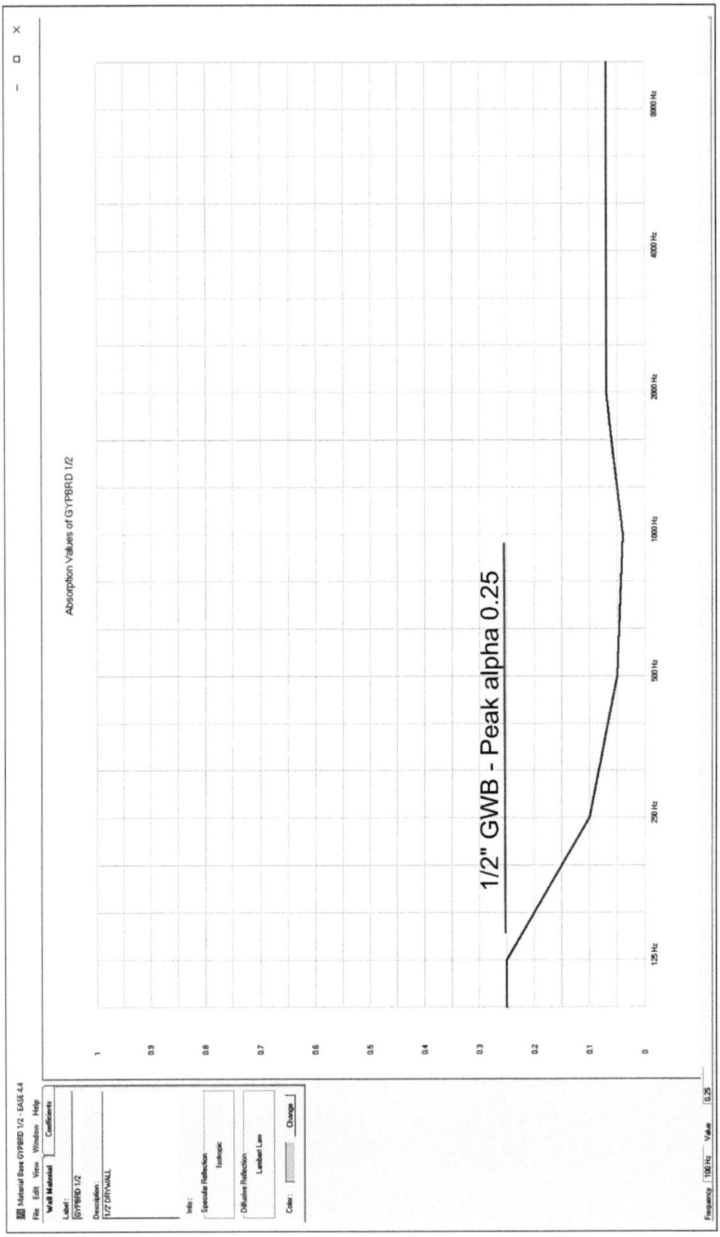

Figure 6.13 Absorption values of ½″ gypsum wall board. The alpha at 125 Hz is only 0.25. At best, this material is only able to absorb 25% per square foot of the LF/VLF energy that strikes it. Also unknown are the stud spacing, well depth, and insulation fill. Each will modify the performance. *Source:* EASE Materials database[5]

5. https://www.afmg.eu/en/ease#features

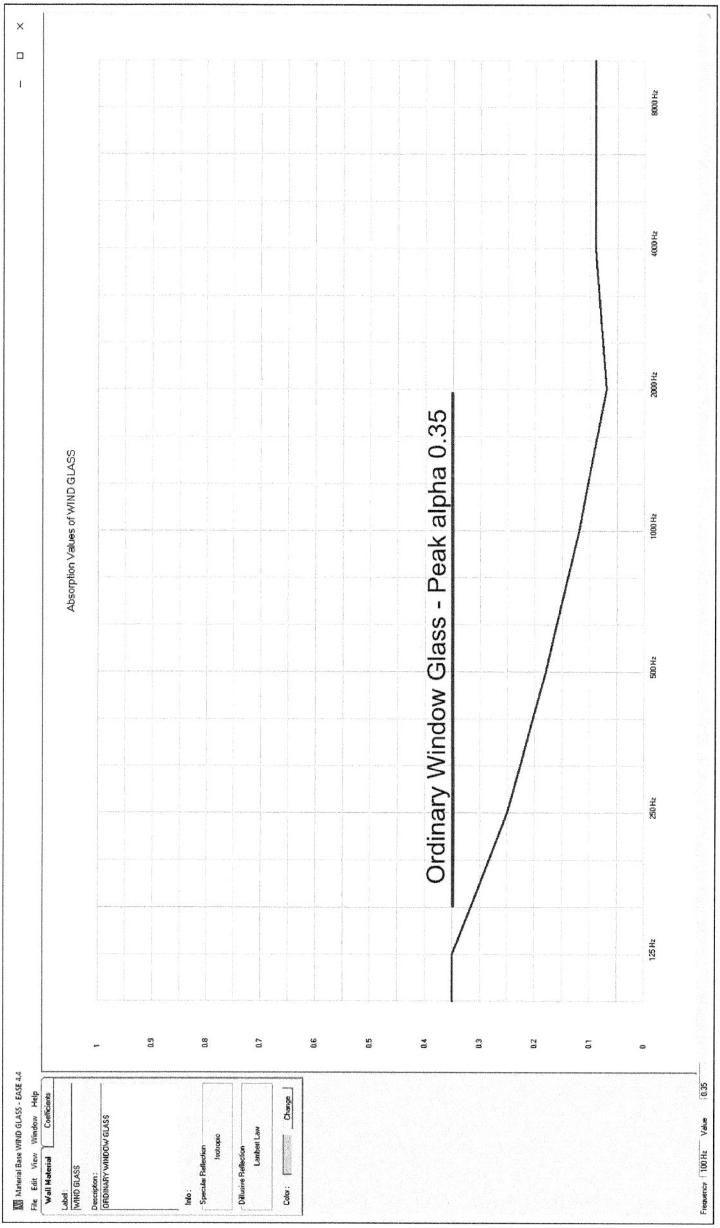

Figure 6.14 Absorption values of ordinary window glass. The alpha at 125 Hz is only 0.35. At best, this material is only able to absorb 35% per square foot of the LF/VLF energy that strikes it. Thicker glass and dual glazing will further reduce the alpha. *Source*: EASE Materials database[5]

I. Helmholtz resonators are a cavity-type resonator designed to sympathetically vibrate within a very narrow frequency range. A common example of a cavity resonator is an empty soda bottle; the air inside vibrates and becomes an audible tone when you blow across the top.

1. The main difference between a frequency-specific Helmholtz resonator and the broadband frequency profile of perf panels and slat absorbers is in the structural captivity of the airspace behind the face.

   a. Helmholtz resonators require a fully sealed airspace cavity behind the front face. Like the empty soda bottle, the specific, narrow-band resonant frequency of that chamber becomes the primary frequency of absorption. Helmholtz resonators are typically specified when trying to tame a single, very specific modal frequency.

   b. Perf panels and slat absorbers have loose airspace tolerances, meaning they should not have a fully sealed cavity behind the front face. In this configuration they perform relatively broadband absorption.

   c. If fully sealed around all sides and back, slat absorbers become a variation of the Helmholtz resonator, but with the advantage of capturing and absorbing a broader range of frequences.

J. The whole can be greater than the sum of the parts:

1. Because we often have a variety of absorption needs, it is possible to design acoustical solutions that have an aggregate effect. To be clear, each treatment is not doing more work than it is capable of, but rather, each treatment location can accomplish the work of multiple tools (see Figure 6.15).

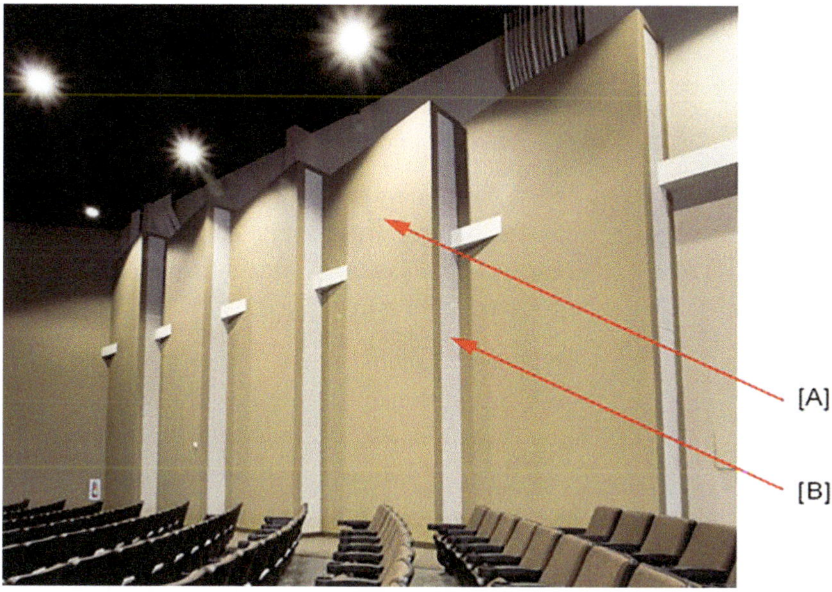

Figure 6.15 Eastlake Church hosts high sound pressure level, contemporary worship services. The two side walls in this 1,600-seat worship center have moderately complex wall treatments that incorporate 2:12 angular rotation to stop side-to-side flutter echo; a little PB absorption via the membrane resonance of the angled sheetrock walls [A]; deep, perf metal absorption cavities behind each of the light colored 16″ return faces of the angled walls [B]; and added modal complexity due to the dimensional variety. The architect (Manuel Oncina Architects) and I worked to create the optimum staggered-wall layout, as well as all other VB and PB treatments the room required. Behind the perf metal panels, the cavities are filled with unfaced, R-30 batt insulation. Because of their size and depth, the cavities provided some additional LF/VLF absorption too. This room has a 0.95 second T_{mid} and an *Optimal* T_{60} Slope Ratio (TSR)* grade. A case study report can be downloaded here.[6]

* This award-winning project was completed and tested roughly ten years prior to the author's formal publication of the TSR theses.

6. https://www.gracenoteds.com/eastlake-community-church

K. Generic architectural finish materials:

1. Commercial-grade carpet is a marginally effective VB absorber. It also has four significant advantages: cost, aesthetics, location, and performance.

   a. Cost: Carpet rarely appears as a line item in the acoustics budget. Therefore, if already planned, it becomes an *acoustical treatment* that adds no additional cost to the owner.

   b. Aesthetics: In almost all cases, there are only two finish options for flooring: hard and soft. Carpet is the ubiquitous soft finish material and is pretty much the only option if a VB absorption product is required. Plus, very few people are offended by the application or appearance of carpet. Well, to be honest, I've been in a few hotels and casinos that had aesthetically offensive carpet—but that's a whole different topic.

   c. Location: Carpet often covers large areas, as well as being evenly distributed throughout a room. Even spot or area applications can be useful.

   d. Performance: While of almost no help in the LF/VLF spectrum, carpet can make a noticeable difference in the range of frequencies that impact speech intelligibility. However, at best, a fully carpeted room will only help reduce some of the worst intelligibility and ambient noise problems. When no other VB absorption is planned, carpet alone should not be considered as an appropriately sufficient acoustical solution.

   e. Note: The absorption properties of carpets vary greatly (see Figure 6.16). The acoustical performance is very dependent on the type of carpet and how it is made.

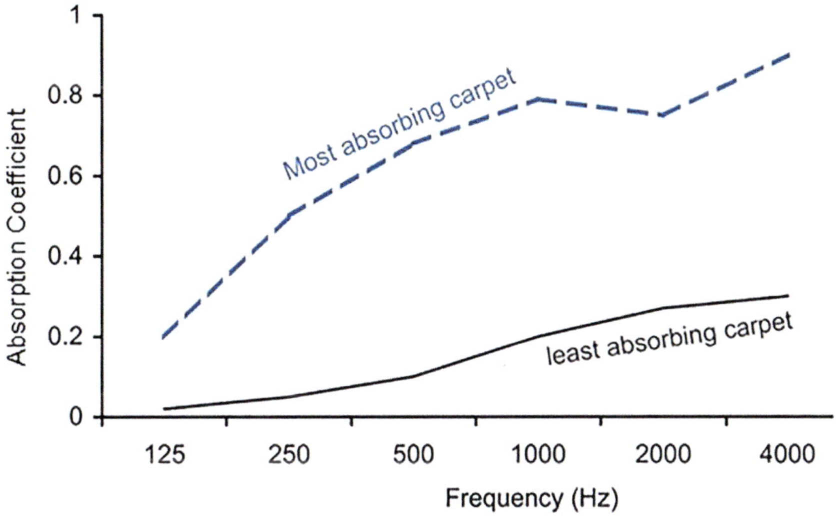

Figure 6.16 This graph shows the least and most absorbing carpets from tables in acoustical publications and engineering textbooks (after TJ Cox and P D'Antonio, Acoustic Absorbers & Diffusers, CRC Press, 2009)

f. While underpadding slightly adds to the performance, it is usually not a feature or expense that is mandatory. However, "to maximize the absorption, the carpet should have an open back with an open cell underlay."[3]

g. When the primary use of a room or venue is geared toward unamplified music, carpet should not be installed in practice/rehearsal rooms, on stages, or in choir lofts. Why? Because most acoustic instruments have significant amounts of their contributory sound energy directed toward the floor. Similarly, carpet is not recommended below seating in worship facilities since vocal sound energy is reduced; thereby negatively affecting the corporate worship experience.

2. Padded seating: Whenever possible, especially in medium and large venues, incorporating padded seating (preferably soft fabric seat and back) should be given serious consideration. Padded seating provides another common finish that greatly helps mitigate one major acoustical problem in some venues.

   a. The problem is the dramatic changes in reverberation times that occur when attendance varies greatly from one event to the next. There is no simple, mechanical, or electroacoustical solution for this scenario.

   b. For most venues, if you must choose one over the other, choose padded seating over carpet. In some cases, doing both is ok too.

3. Fabrics: The phrase *acoustically transparent* is used to describe fabrics that do not alter, impede, or absorb sound energy. When used as a finish covering for the various sound absorption materials previously listed, it is critical to specify acoustically transparent (AT) fabrics. Guilford of Maine[7] offers one of the widest selections of AT fabrics.

   a. If you need to specify fabrics for aesthetic reasons, use AT type so the material does not contribute acoustically. Also, AT fabrics are not meant to be draped.

   b. If you want or need to apply fabrics for their absorptive properties, it is usually best to use heavy (14 to 18 oz. per sq. yd. or greater) velour-type materials, gathered to half area.

L. There is one universal variable to the effectiveness of all VB and PB absorption products: the thickness or depth of the

7. https://www.guilfordofmaine.com/acoustic

treatment. Thicker, deeper materials generally absorb more sound and reach down to lower cutoff frequencies. But more is not always the goal. Too much or the wrong type of treatment can be a problem, too. For more on this topic, see the *Parametric Acoustics*[8] white paper.

6.4 Diffusion: The physical process of diffusion causes reflected sound energy to be scattered in a room. Our detailed understanding of this phenomenon is only about forty years old. The theories used today were developed by Dr. Manfred Schroeder[1], circa 1970, but weren't codified and brought to market for the first time until the early 1980s.

A. You have no doubt noticed that the words diffusion and scattering have been used throughout this chapter. Diffusion and scattering are closely related subjects, but their net results are somewhat different. The details of the differences go beyond the basic needs of this commentary but here are the key points:

1. When sound waves strike rigid convex or geometrically complex structures (see Figures 6.17, 6.18, and 6.19), they burst into countless smaller reflections, which are then randomly *scattered* in many directions. Visualize a beam of light that's scattered when it strikes a disco ball.

2. Random scattering implies no specific spatial or temporal controls have been implemented.

3. Common architectural elements that scatter incident sound are stairs, stage lips, chairs, pews, cylindrical pillars, balconies, soffits, statuary, and all other complex geometric elements and curved surfaces that present in a convex manner.

4. Bottom line: Architectural elements that scatter sound are highly desirable and rarely problematic.

---

8. https://www.prosoundweb.com/parametric-acoustics-the-case-for-the-parametric-method-of-acoustic-treatment-second-edition/

Figure 6.17 This is an RPG Planarform™ Wrinkle G tile. This type of pseudo-random texture scatters sound. *Source*: RPG Acoustical Systems, LLC®

B. While round columns and other convex structural features make excellent scattering structures, this is not true of concave structures. Once again, convex features scatter sound; concave features *focus* sound—often detrimentally. Bottom line: Architectural features that focus sound are rarely desirable.

C. When sound waves strike custom-built diffusion modules (spelled diffusors or diffusers), they are scattered in fairly even and predictable directions. Size, depth, and geometric topology all factor into the various types of custom diffusers. Figure 6.20 shows an example of a quadratic residue diffuser (QRD). QRD designs are based on the mathematical number-theory sequences developed by Manfred Schroeder.[1]

   1. Much of the diffused sound energy remains *alive* and actively moving through the room. However, small amounts of energy also get trapped and absorbed in the complex geometry of the diffusive boundary.

Figures 6.18 and 6.19 GFRG (glass fiber reinforced gypsum) is a *thin cast* alpha gypsum product. These can be factory molded into virtually any shape or size. Products such as these allow architects and designers to create unique scattering shapes and designs that would not be possible with traditional materials. *Source:* Shutterstock.

Figure 6.20 A QRD-formatted diffuser. This is the Diffuser 73 from Shubert Studio Akustik. The primary range of frequencies controlled by this 3-unit module is 400 Hz–5 kHz. *Source*: shubertstuio.eu

2. Diffusion is a frequency-specific treatment methodology. The larger the wells and cavities in a diffuser panel, the lower the frequency range that is effectively diffused. Smaller wells and cavities are used to better diffuse and scatter at higher frequencies.

3. When properly defined and implemented, diffusion is a highly desirable treatment tool. Architectural features and finishes that diffuse and scatter sound provide a uniquely beneficial hybrid of sound control. They break up problematic slap and flutter echoes and absorb a small amount of the energy that's being scattered, while also allowing a reasonable amount of reverberant energy to be maintained.

D. LF and VLF scattering is accomplished by deploying large, moderately complex geometrical shapes. Figure 6.21 shows a custom ceiling treatment that is dimensionally large enough to provide broadband scattering.

Figure 6.21 Note the complex, convex geometry of the ceiling in the St. Callistus Catholic Church sanctuary. Rather than having a large, flat, hard ceiling, the architect (Andrade Architects) adapted my recommendations for "something more complex." Together with the stained concrete floor and padded pews, the acoustical character of this room was wonderful, and worked well for both traditional and contemporary worship, regardless of attendance. This ceiling treatment is a mirror-image on the opposite side of the room.

1. The wavelength of a 100 Hz tone (approx. fundamental of a baritone voice) is a little over 11′. A low bass guitar note (50 Hz) is double that wavelength. Because the wavelengths are so long in this region, LF or VLF scattering requires treatment that is many times larger than what is needed for MF or HF diffusion or scattering.
   a. It is not necessary to construct full-wavelength LF or VLF scattering elements. However, at least one-quarter of the target wavelength is needed.

b. A convex structure that is at least 5.5′ in its smallest dimension can provide useful scattering for frequencies of 50 Hz and above. A solid stage with a 3′ elevation will provide a little scattering down to about 95 Hz.

2. For most public facilities, large building features or structures are often the only practical application of LF/VLF scattering. For some rooms, large structural protuberances or ceiling clouds can serve this purpose. In others, providing there is sufficient mass, balconies, pews, chairs, and stairways can each contribute to LF/VLF scattering.

3. In summary, diffusion/scattering elements are very effective tools when properly applied. They can also be more expensive to implement than either of the other two tools.

E. Note: While acoustical scattering and diffusion are desirable outcomes, employing loudspeakers that randomly propagate scattered sound energy is not.

6.5 Surface area and location challenges:

A. Regardless of the acoustical tools or materials used, installing the appropriate amount of each material is critical. In their book *Acoustic Absorbers and Diffusers*, world renowned acousticians Trevor Cox and Peter D'Antonio summarize it this way, "Good acoustical design results from an appropriate combination of absorptive, reflective, and diffusive surfaces."[9]

---

9. Acoustic Absorbers and Diffusers: Theory, Design and Application. Trevor Cox and Peter D'Antonio. https://www.taylorfrancis.com/books/mono/10.1201/9781315369211/acoustic-absorbers-diffusers-trevor-cox-peter-antonio

B. The successful application of any new or custom treatment is based on the total surface area of each material or product and the locations in which they are placed. The formula used to calculate a room's T_{60} relies heavily on the total surface area of each finish material applied. However, grouping the materials together on one wall or in one confined area is usually not fully productive, recommended, or cost-effective.

C. General and/or specialized acoustical treatment(s) should be considered under these conditions:

1. Good to excellent speech intelligibility is a clearly defined deliverable.
2. Ambient background noise is, or may become, a problem.
3. Live, acoustic, and/or amplified music will be performed.
4. When characterizing the most demanding uses for a room, you believe the whole room may need to be adjusted to a more appropriate T_{mid} and/or TSR. An obvious example occurs when an old, traditional church sanctuary is being converted (modernized) to accommodate today's highly amplified contemporary worship music.
5. Acoustical zones or spot treatments are required.
   a. There are situations that may require a small, dedicated, acoustical zone that differs from the overall room requirement, such as a choir loft or drum kit enclosure.
   b. There are also scenarios that call for a spot treatment with absorptive or diffusive materials (see Figure 6.22) in order to mitigate a localized problem in an otherwise well-behaved room.

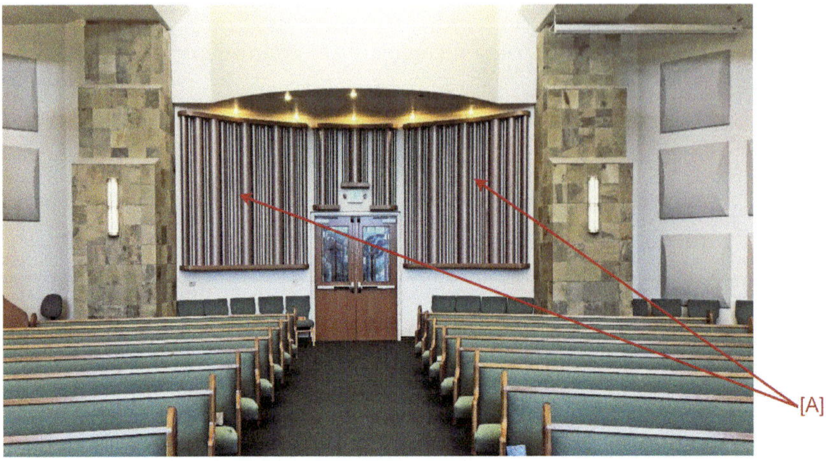

Figure 6.22 At Vista United Methodist Church, an architectural "feature" in their sanctuary—the concave entry wall—caused sound energy to collect and reflect back to the last several rows of pews. Depending on where one was sitting, the concentrated reflections arrived between about 50 to 100 ms after the program audio, causing a strong, audible, slap echo. The solution was a custom PVC pipe diffuser wall [A], using 2.5″, 4″, 6″, and 8″ PVC tubes that conformed to the unique, concaved curvature of the wall. Custom paint was selected to match other existing interior finishes. The tubes also blended into the traditional pipe organ motif in this room.

CHAPTER 7

WHERE DOES ALL THE UNUSED SOUND GO?

7.1 Once the initial sounds get started and move into a room and past the ears of the audience, do they go on forever? If not, why not? The answer is no; they don't last forever. The *why not* takes some explanation.

7.2 The basic *law of conservation of energy* states that "energy can be transformed, but never created or destroyed." This means that all the energy a sound source puts into a room must be accounted for by either absorption (turning sound energy into heat), transmission (passing through the enclosure to another space), or reflection (adding to the sound level).[1]

A. If sound energy was to build up in a completely sealed, totally reflective room, and the sound source continued to put out sound energy, none of the energy would be absorbed. "Over time, we could theoretically arrive at a pressure that would be explosive."[1]

7.3 Presuming an enclosed space, when sound energy reaches any absorptive or diffusive/scattering boundary, barrier, object, or person—regardless of size—some or much of its energy is converted to heat. The remaining energy is either reflected and/or scattered, at which time it continues along its new path(s) until it strikes other objects or is otherwise transmitted out. This process

1. *Sound System Engineering*, Second Edition. Don and Carolyn Davis. Howard Sams & Co.

continues until the remaining sound energy becomes inaudible, then completely undetectable.

A. The physics: In simplistic terms, friction is the mechanism that converts sound energy into heat. Heat is inaudible. This all happens on extremely small human scales. You can't see the movement or feel the heat, but it is there.

B. Even without absorptive and/or diffusive materials, sound energy is also depleted by the friction that exists between air molecules as they are pushed and pulled through the air. Air absorption consumes high frequency (HF) energy the most, and as the frequencies move down the spectrum from mid frequency (MF) through very low frequency (VLF) energy, they are absorbed less and less. This explains why, in most rooms, finding excess reverberation above 2 kHz is rarely a problem; the air in the room soaks it up.

1. Along with the mass of the walls, doors, windows, floor, and/or ceiling, it is also one of the reasons why you often hear low frequency (LF)/VLF energy from the neighbor's party in significantly greater amounts than the MF/HF sounds.

7.4 Another factor of energy depletion comes via sound transmission. This occurs when sound—either acoustically or structurally—passes through walls, doors, windows, floors, and ceilings.

A. The construction materials and techniques used in most buildings allow a certain amount of sound to escape the structural envelope. This is generally not by design, nor is it necessarily a good thing, however it is one more way that sound energy is depleted.

7.5 What about outdoor sound?

A. Air, various natural and man-made structures, and distance all play significant roles in the absorption, reflection, diffraction, and scattering of sounds as they propagate outdoors. Primarily, there are six controlling factors:

1. The *inverse square law*[2] (see Figure 7.1) explains the dissipation of sound pressure at the rate of 6 decibels (dB) per each doubling of distance. The initial wave of energy spreads out more and more as it moves away from its source. Therefore, less and less energy is audible at various points along its flightpath.

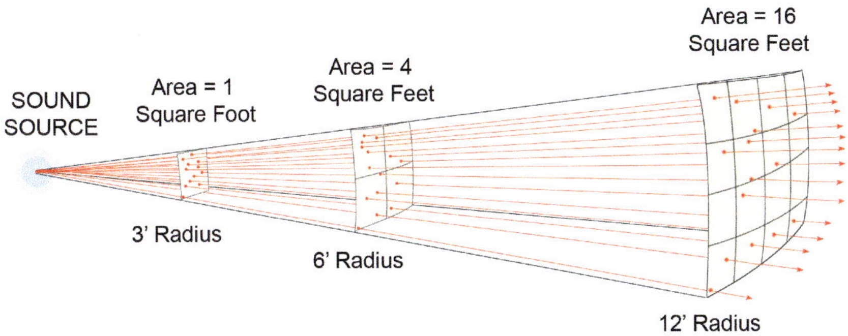

Figure 7.1 This graphic gives a visual representation of the inverse square law at work. Simply stated, the sound pressure of a spherically expanding wavefront decreases by 6 dB for each doubling of the distance from its source. Graphic by Atiq Ahmed[3]

2. Depending on the surrounding structural and environmental conditions, sound energy may encounter any number of boundaries and/or resonant objects. Such items further dissipate sound energy when contacted.

3. Ground reflections and/or absorption depend on what substance(s) covers the surface. Though minimal, grass does absorb some MF and HF sound. On the other hand, water is primarily a reflector of most frequencies.

2. https://www.acoustic-glossary.co.uk/definitions-i.htm

3. https://www.atiqahmed.us

4. Standing or seated audiences provide measurable amounts of ground level MF and HF absorption.

5. Generally speaking, there are no room modes outside. The resonant peaks of modal energy found inside a room do not build up and linger over time.

6. Some additional losses occur—and vary—based on wind direction, temperature, and humidity. Eventually, the sound pressure is sufficiently reduced to a point where it falls below our threshold of hearing.

B. Outdoors, LF and VLF frequencies also follow the inverse square law, but generally, they maintain audibility over greater distances. There are at least three controlling factors for this behavior:

1. Air absorption is less and less efficient as frequencies decrease and wavelengths get longer. This explains why LF and VLF sounds seem to travel further before being inaudible.

2. Ground reflections are typical because the LF and VLF wavelengths are so large, and there is rarely much absorption achieved by most ground covers.

3. And once again, some amount of additional loss occurs in varying amounts based on wind direction, temperature, and humidity. Eventually, the LF and VLF sound energy is sufficiently reduced to a point where it falls below our threshold of hearing.

CHAPTER 8

ROOM GEOMETRY—THE GOOD, THE BAD, AND THE UGLY

8.1 Architectural geometry and architectural acoustics are inextricably linked. Most rooms that have complex interior geometry will complement and support *good sound* better than ones with simple geometry. This chapter attempts to explain why *room geometry* has such a significant impact on sound and acoustics.

8.2 The importance of room geometry all boils down to issues related to flutter echo (see Chapter 5, Section 5.3) and room modes (Chapter 5, Section 5.5). Given enough volume and reflective surface area, reverberation can and will exist in all rooms, regardless of shape.

A. The ubiquitous six-sided, rectangular *shoebox* room is a very common architectural shape. Each opposing pair of surfaces are parallel. All corners are set at 90-degree angles. It is a very simple three-dimensional shape and is typically very cost-effective to build.

B. When more complex interior geometrics are suggested, we are talking about rooms with more than six major surfaces, convex curves, obtuse angles, tilted or rotated walls, alcoves, balconies, ceilings that are not parallel to the floor, sloped or stepped floors, etc.

C. In addition to the basic geometrical decisions, other important factors to consider are elements such as the physical placement of the sound source(s) in relation to the listeners, a room's interior dimensions and symmetry, the inclusion and use of a sound reinforcement system, and the low frequency (LF)/very low frequency (VLF) content to be presented.

D. Are you being asked to design a 30′ × 30′ classroom, a huddle space, a worship center, a multipurpose building, a sports arena, or a performing arts venue? Acoustically, each has a unique set of challenges.

E. Let's now consider the floor plan of various room shapes and volumetric ratios. The following list includes many possibilities. Acoustically, not all are particularly good, but they do illustrate key concepts relating to the ideas being expressed throughout this commentary. Following the listing of common geometries, beginning with spherical rooms, an overview of common ceiling topologies is outlined in Section 8.17.

8.3 Spherical rooms

A. The inside of a sphere (see Figure 8.1) is generally the worst acoustical shape. Because this isn't a practical building shape for most public venues, it is rarely seen and not worth much further commentary other than these few points: The problems with spherical or near-spherical rooms are primarily modal and focused, specular reflections.

1. Spheres simply do not provide enough modal density and variety. There is only one interior dimension: the diameter of the sphere. A perfect sphere produces only one, dominant, resonant frequency.

2. Much like concave walls and domes—which are covered in more detail later on—the untreated interior surface of a sphere causes reflected energy to concentrate in a very specific area of the room. That area is at and near the focal point of the spherical radius. This concentration of sound energy is often much louder than the nominal sound levels in other parts of the venue and may also include audible flutter echoes.

Figure 8.1 The 18,600 seat MSG Sphere in Las Vegas, NV is currently the most iconic spherical building in the world. It relies heavily on very specific and expensive audio system propagation and absorptive acoustical treatments. The immersive experience of sight and sound in this venue is simply amazing to experience. However, if you're considering this geometry, be prepared to add one or more zeros to the construction budget.

3. For near-spherical rooms, there can be multiple focal points. The size of the room and the arc radius of the various curves determine their location.

B. There is also the phenomenon of *whispering galleries* to consider. This is the experience that occurs when you whisper near the wall and some other person, far away, hears you loud and clear.

C. If you are planning for a spherical room that will be used for public assembly, one of your first phone calls needs to be to an acoustical consultant. There will be much work and expense required to make this shape sonically tolerable.

D. Conversely, the outside of a sphere is an excellent shape when located inside of almost any room. Regardless of size, it becomes a very effective convex scattering element.

8.4 Cubic rooms

A. Cubes are the worst of the realistically viable room shapes. The problem is that all three major dimensions are equal, and each of the three pairs of surfaces are parallel. This configuration results in a bad set of room ratios: length to width, length to height, and width to height. In short, due to a lack of dimensional variety, there is not enough modal density and variety. See more on room ratios in Section 8.14.

B. Cubical rooms have fewer and stronger modal peaks and valleys than rooms with optimal aspect ratios. Also, if not carefully treated on at least three non-opposing faces, problematic flutter echo will potentially dominate the listening experience. Some careful planning, and a reasonable treatment budget will be required to make a cubic room sonically viable.

8.5 Round/cylindrical rooms

A. Round rooms—with or without a parallel floor and ceiling—are a difficult shape for any sound-related activity (see Figure 8.2).

B. Acoustically, round rooms behave much like spheres. Without significant amounts of absorptive and/or diffusive materials being incorporated into the curved walls, this footprint will likely be challenging for public assembly of any size. Ovular and egg-shaped rooms have similar problems.

C. The primary issue with cylindrical and oval rooms is the focusing of excess sound energy toward a fairly small area, at and around the center of the arc radius. You should be starting to notice a theme: Regardless of size, concave boundaries are not acoustically friendly for most rooms used for public assembly and/or any kind of audio presentation.

D. If you must design a round room, moving to a cone shape (see Figure 8.3) will help, providing the tilt gradient is at least 1:12 and the floor and ceiling are not parallel.

Figure 8.2 A small, glass, cylindrical meeting room. Looks cool! Sound . . . not so much.

Figure 8.3 A large, cone-shaped building.

8.6 Rectangular rooms

A. Rectangle plans are probably the most common shape for commercial buildings. The geometry is simple and obvious. Picture the common shoebox. Once the decision has been made to build a rectangular room, the subject of room ratios should become a priority. It is prudent to evaluate and potentially adjust the ratios if the space will be sound critical and used to generate or receive any meaningful audio content—be it spoken word or music.

B. The length, width, and height dimensions should all be different and not an exact multiple of each other. For example, a rectangular room that is 12′ high, 24′ wide, and 36 or 48′ long would *not* be a good place to start.

1. Direct multiples are problematic because they lack dimensional variety. A poor distribution of room modes will occur. Because of this, significantly large gaps will occur between the few modes that develop. The result is that the most prominent modal resonances can be easily heard by most people. Ideally, room modes should be spread out and spaced as evenly as possible.

2. In case this wasn't made clear earlier, the goal is not to eliminate all modal activity. What is desired is a significantly large number of closely and evenly spaced modal resonances—at and below F_S.

C. Contrary to common belief, flat, parallel walls are not automatically a negative acoustical factor. However, there are potential acoustical problems if the room is not designed with some care beyond the basic construction requirements.

1. The 19th-century European concert halls all have parallel side walls. They provide psychoacoustic widening of the natural sources from the stage end, and their ornate decorations provide some useful scattering.

2. Parallel bare walls will very likely need some form of acoustical treatment on at least two non-opposing walls. If left untreated, flat, reflective, parallel walls create an environment that generates and supports flutter echo.

3. If the floor and ceiling are parallel, some form of acoustical treatment will likely be needed on at least one of those surfaces too. In small and medium-sized rooms, this is often carpet and/or acoustical ceiling tiles. However, because it is so thin, carpet is one of the least effective velocity-based (VB) absorbers.

8.7 Triangular rooms

A. A triangle is not a common building shape, but it does present one nice feature: the side walls are about as far from parallel as possible. Flutter echo shouldn't be a problem, unless it occurs between the floor and ceiling.

B. To better address appropriate room ratios, try to avoid equilateral triangle plans.

8.8 Quarter- and half-round rooms

A. In recent years, these shapes have become popular for performing arts and house of worship venues. While these shapes can be quite good when considering sight lines and a need to bring the audience as close to the stage as possible, they, too, can present acoustical challenges.

1. One example is that the side walls will not provide any useful, early reflections. Side wall early reflections are an important feature in rooms that are designed specifically for unamplified speech or acoustic music. Once a sound reinforcement system is employed, the need for early side wall reflections often becomes moot.

B. Any concaved interior wall surface presents an acoustical paradox.

1. If the concave surface is located at or near the rear wall of a stage or platform, it may serve as an effective and perhaps even desirable acoustical reflector that may enhance certain styles of unamplified speech or musical performance. In theatrical terms, the part of a stage closest to the rear wall of the stage is called *upstage*.
2. If the rear wall of the *house* (opposite the stage) has a concave shape (see Figure 8.4), it will be acoustically problematic and will very likely need some specific treatment (see Figure 8.5) to minimize the negative effects (arc radii focusing) of the reflected energy bounding off the curved surface.
3. Regardless of whether the back wall is a true curve or a faceted curve, the detrimental effects are very similar.

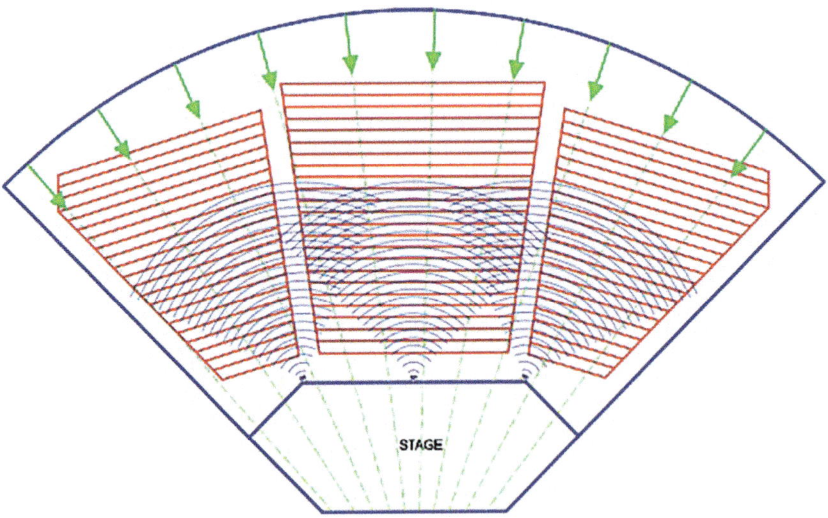

Figure 8.4 If you have been following along for a while now, you should be able to easily spot the main problem with this quarter-round, fan-shaped room. This shows an over-simplified view of how sound propagates out of three loudspeakers and reflects off the concaved rear wall toward the upstage wall. Not only does the audience hear the direct sound, but also the reflected sound at various, offset, timing intervals.

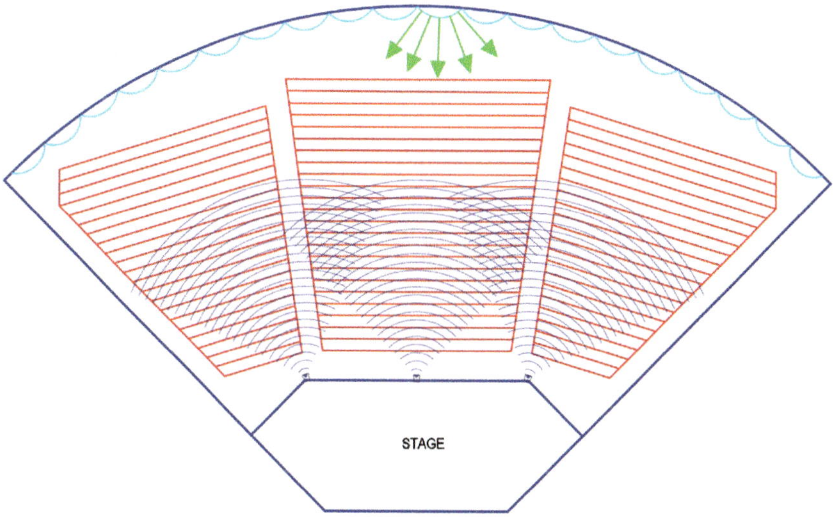

Figure 8.5 One solution used to counteract the issues with a concave wall is to incorporate lots of smaller convexities. Each outthrust discretely scatters the oncoming energy that strikes it. If needed, varying the radius of some scheme elements will improve LF performance. Using a saw-tooth pattern will also work, but with less efficiency.

8.9 Trapezoidal rooms

A. Like the quarter-round shape, a trapezoidal room (see Figure 8.6) has one notable advantage: at least one set of walls is not parallel. This shape helps break up side-to-side flutter echoes and creates more complex modal behavior in the LF and VLF regions.

   1. The severity of the side wall angles determines how much of the center seating is removed from useful side wall reflections and pushed toward the edges.

B. The greatest disadvantage of the trapezoid shape is that the front and back walls are often parallel. This configuration will require further evaluation and treatment if not properly factored into the initial design. Here are some solutions to consider:

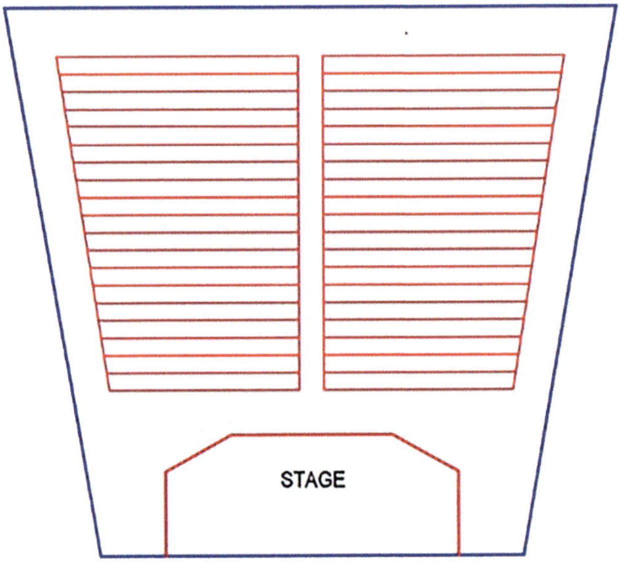

Figure 8.6 This basic trapezoidal plan is problematic if the back and upstage walls aren't tilted, staggered, or sufficiently covered with VB absorptive treatments.

1. Tilt one or both parallel walls using the minimum 1:12 slope gradient offset.
2. Add convexities, as shown in Figure 8.5, to one or both walls.
3. Add perforated or slat absorption panels to cover at least 50% of one of the surfaces—preferably the back wall, opposite the stage.
4. Include a nonparallel slope to the floor and/or ceiling.
5. If needed, mix in some pressure-based absorption devices along with the VB treatments.

8.10 Pentagonal rooms

A. The pentagon (see Figure 8.7) is my favorite flat-wall shape because it offers a large seating area, good sight lines, and no major parallel or concave walls.

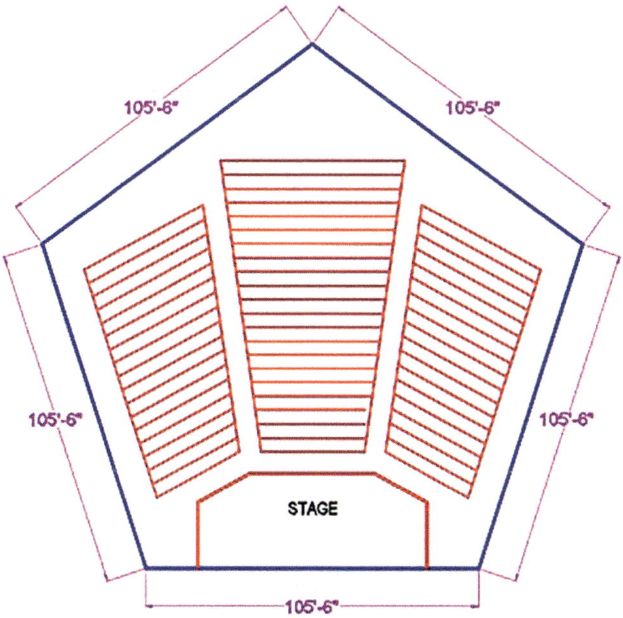

Figure 8.7 Pentagon-shaped rooms are an excellent place to start. Regardless (within reason) of how you dimensionally vary the five walls, you should never have flutter echo. However, this example uses equal-length walls, so the room ratios are less than ideal.

B. As nice as this shape may appear at first glance, there are still potential acoustical issues to be considered for this or any other room shape. These are based on:

1. The width of each wall segment, which translates into finding the best possible combination of room ratios (see Figure 8.8).
2. The height and slope angle(s) of the ceiling.
3. The slope angle of the floor.
4. The location of various major structural elements, such as a stage.

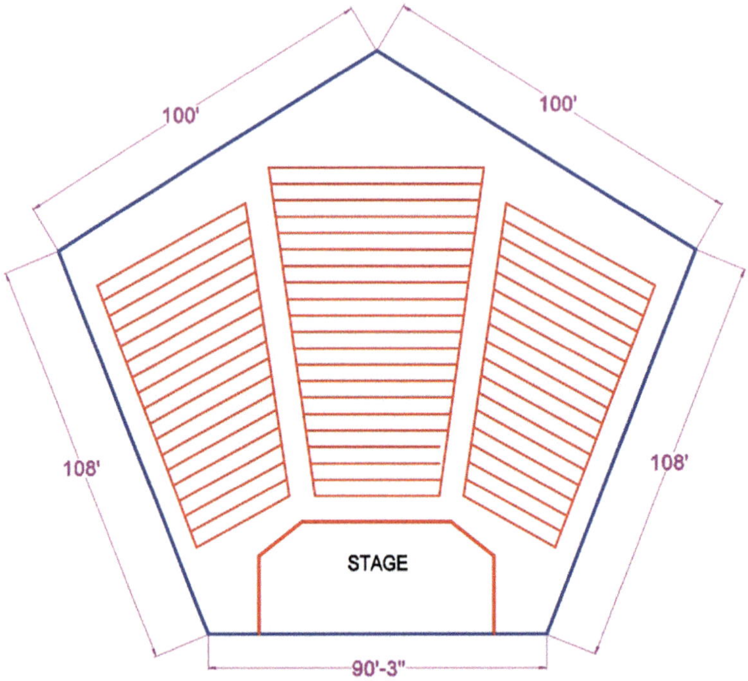

Figure 8.8 Notice the subtle dimensional shift between the five major walls. This figure provides some much-needed room-ratio variety—notice the beneficial modal complexity.

C. Presuming a ceiling height of 30′ and that the floor and ceiling are parallel, the room ratio for Figure 8.7 is 1:3.52. Because all five sides are of equal length, this is not an optimal layout.

1. After minor dimensional adjustments, the ratios shown in Figure 8.8 become more advantageous—1:3.01:3.33:3.60.

2. To further improve the pentagon, vary the length of each wall segment so no two are equal and vary the slope of the floor and/or ceiling so they are not parallel.

8.11 Hexagonal rooms

A. Adding more wall facets is not necessarily better. The hexagon plan (see Figure 8.9) falls back into the shortcomings of the rectangular group because it has multiple parallel walls. All the same precautions must be taken related to flutter echoes, room modes, and room ratios.

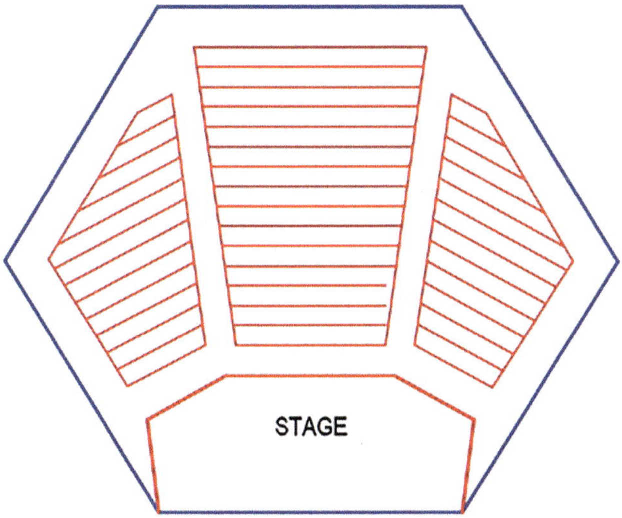

Figure 8.9 This graph shows an equilateral hexagon. If a hexagon plan is unavoidable, at least try to make each wall a different length, and include a sloping floor and/or ceiling. Dimensionally, a 5 to 10% variation is sufficient.

8.12 Heptagonal and octagonal rooms

A. Much like the pentagon, heptagons (seven sides) are a potentially good shape to consider. However, construction costs will now start climbing, but with little or no acoustical benefit.

B. Octagonal (eight sides) plans have the same issues as hexagon and heptagon plans.

C. In both cases, as the number of facets increases beyond six, rooms become closer and closer to being round, which is not a desirable acoustical goal.

8.13 Concave and convex surfaces

A. There is no question that architectural structures take on many other more complex shapes than those previously described. Acoustically, all solid architectural finishes constitute either flat or curved surfaces. Let's now look at the merits of these two curved-surface shapes.

B. Concave: All concave surfaces are acoustically challenging if finished with hard, reflective materials such as sheetrock, concrete, wood, glass, etc.

1. As noted earlier, concave surfaces concentrate reflected sound energy into a small area. The density of concentration and the size of the focal area are largely determined by the arc radius of the surface, the distance between the focal point and the listener(s), and the overall size of the curved surface.

2. Focusing reflected sound is usually not desirable and should be avoided in most situations. Yes, there are exceptions to this guideline, but they are few.

3. Faceted convex planes are also problematic when there is a large number of facets that result in too much of a cylindrical profile.

C. Convex planes: All convex planes bring the exact opposite results, and are generally encouraged, where and whenever possible.

1. Convex planes are inherently diffusive, which is good. More often than not, convex planes result in improved acoustics and sound management.

2. Round columns of all sizes are considered convex planes.

3. Faceted convex planes are less complementary but still beneficial. The number of facets and the arc radius of the overall curvature will determine the relative benefits. More facets generally bring better, more even results.

4. If a building's design can include two or more large, mirror-image, convex planes, regardless of location, the need for additional diffusive or absorptive materials may be reduced.

8.14 Room ratios

A. Dimensionally, rectangular rooms are commonly described as having a given height, width, and depth. However, the nouns length, width, and height are also commonly used and can be considered equal and synonymous.

B. In rectangular rooms, the ratios start with the smallest of the three dimensions representing a factor of one (1). In the 12′ × 24′ × 48′ example noted in Section 8.6.B, 12′ equals 1. From there, each of the other dimensions is compared to provide its ratio—relative to 1. Therefore, the room ratios for this example are 2.0 and 4.0, which are commonly presented in ratio sets like this: 1:2.0:4.0.

C. In recent research on room ratios in small rectangular rooms, it has been found that "the length-width ratio is much more important than the width-height ratio. The length-width ratio should be within 1.15–1.45. The height can be chosen more freely without compromising the acoustical quality."[1]

D. There are many *optimum* ratio sets stated in the technical references. For example, 1:1.4:1.9.

   1. A roughly 9,000 ft^3 room might have a 15′ ceiling, be 21′ wide, and 28.5′ deep. This would be a very nice rectangular room shape to work with. Need a bigger room?

---

1. *Preferred Dimension Ratios of Small Rectangular Rooms.* Jens Holger Rindel Odeon A/S, DTU Science Park, DK-2800 Kongens Ly.

Double all these dimensions so the room becomes 42′ × 57′, with a 30′ ceiling. The ratio set stays the same.

2. Many other good-to-optimum ratios exist, so you'll have several options to consider. Your friendly, local acoustician can help you find the best ratio set for your project.

3. If the room you are designing will be something other than a rectangle, the ratios are somewhat less critical because you are potentially introducing much more modal variety. Again, having lots of modal variety is a good thing.

4. Designing around one of the optimum ratios is not the only factor to consider, but it is a good place to start if you are designing a rectangular room that will be used for some type of sound generation or reception.

8.15 Symmetry

A. If the purpose of a room is to support sound-critical live performances or the presentation of other meaningful, audible information, it is tremendously helpful to design a symmetrical interior shell and seating layout. To be clear—symmetry does not imply simplicity and complex geometry is usually a beneficial acoustical feature.

B. One of the prime goals and challenges for sound system designers is to deliver even sound coverage to all seating areas. Architects can help with this. There are significant performance, cost, and aesthetic benefits achieved when a room is designed with mirror-image symmetry along the center line axis of the stage or platform. It is important to note the following points:

1. Good quality sound reinforcement requires careful time-domain implementation. A sound system that is not properly *time-aligned* is analogous to a photograph that is out of focus.

2. One critical architectural deliverable is providing the best possible sight lines. Delivering good-to-excellent amplified sound propagation is analogous to supplying good-to-excellent sight lines.

3. If one of the design directives is to have multiple terraced floor elevations on the main floor, try to limit the variations to be no greater than +/– 2′.

4. Every major building element or feature that introduces significant asymmetry, obstructions, or other irregularities to the floor plan, ceiling, and/or audience seating area(s), may result in an inconsistent acoustical environment.

    a. Additional sound devices, labor, structural points, and aesthetic anomalies can all become unexpected challenges brought on by the need to support an asymmetrical seating and/or building structure. Obviously, each of these adds cost and technical complexity to any project.

5. There are some exceptions—times when specifying *zoned acoustics* is an appropriate and desirable goal. Application-specific acoustical zones need to be carefully planned and purpose-built.

8.16 Flex spaces

A. Occasionally, multipurpose rooms are used with various seating configurations where the *front* or focal point of the room may be located just about anywhere. These are sometimes called flex spaces.

    1. The Perelman Performing Arts Center in New York City is a good example. According to Rob Miller at Threshold Acoustics,[2] "the acoustic challenges were resolved with largely diffusive surfaces."

2. https://www.thresholdacoustics.com/

B. In rooms with simple geometry, the acoustics—be they good, fair, or bad—will not change because of the orientation of the seating layout. However, flex rooms do present additional sound, video, and lighting system costs along with operational complexities.

8.17 Ceiling layouts

A. Ceiling layouts are another important element within the acoustical environment. Ceiling symmetry is just as important as floor plan and wall symmetry.

B. If a room is to have a single interior ceiling slope, it is helpful to have the highest point in the room located above the stage or platform, with the ceiling gradient sloping downward as it moves toward the opposite wall of the room.

1. The slope gradients should be in the range of 1:12 to 3:12. Anything that is less than 1:12 is acoustically insignificant and may as well be flat. A gradient that is greater than about 3:12 has its challenges too.

2. If a sloped or stepped floor is planned, slant the floor and ceiling in opposite directions, so the angle opens widest at or near the front edge of the stage. In theatrical terms, the front of the stage is called *downstage*.

3. Gradual, yet acoustically significant slope changes are much preferred over dramatic changes. One challenging example can be seen in buildings that have a central clerestory, with a significantly lower ceiling profile beyond. See more on this in Section 8.18.

C. The A-framed roof structure: If a room design calls for an A-frame ridgeline, it is best if the ridge beam runs parallel with the center line of the stage or platform—not perpendicular.

1. Several presentation and propagation challenges are created when the ridgeline is turned sideways so it is parallel with the downstage edge of the stage. This configuration tends to encroach on appropriate loudspeaker placement and the size and location of video screens or displays.

2. It is also best when the overall slope gradient on either side of the ridgeline is *flattened* so the lowest point of the slope at the outer walls is equal to or higher than about 50% of the peak ceiling height. Read a slope gradient of 6:12 or less. For example, a room with a 30′ ridgeline should have walls that terminate at 15′ or more in height.

3. For ceilings that slope continuously from one side to the other, the 6:12 recommendation applies too, though a 3:12 or 4:12 gradient would be even better.

8.18 Clerestory roof structures

A. In rooms that have uniform shapes and ceiling height, reverberation will behave and sound the same in every seat. In rooms with clerestory roof structures, reverberation increases directly below the clerestory (see Figure 8.10).

B. Because reverberation time is directly related to room volume, the sudden expansion in volume under the clerestory creates reverberation times that are somewhat longer than those measured in seating areas on either side. The length, width, and height of the clerestory control the variation.

C. To help normalize the reverberation under a clerestory, a good solution is to add VB absorption treatment where possible. Figure 8.11 shows three different options.

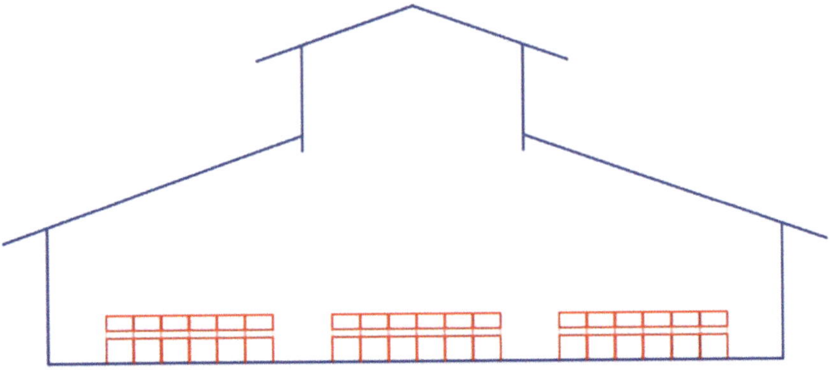

Figure 8.10 Structures like this section of a central clerestory present with two different acoustical zones. Without custom VB absorption installed in the clerestory zone, people seated in the center of the room will experience more reverberation and worse speech intelligibility and music clarity. Those seated in the left and right wings will usually experience less reverberation as well as better intelligibility and clarity, even if the clerestory is left untreated. If the clerestory is properly treated, listeners will experience very little acoustical difference, regardless of where they sit.

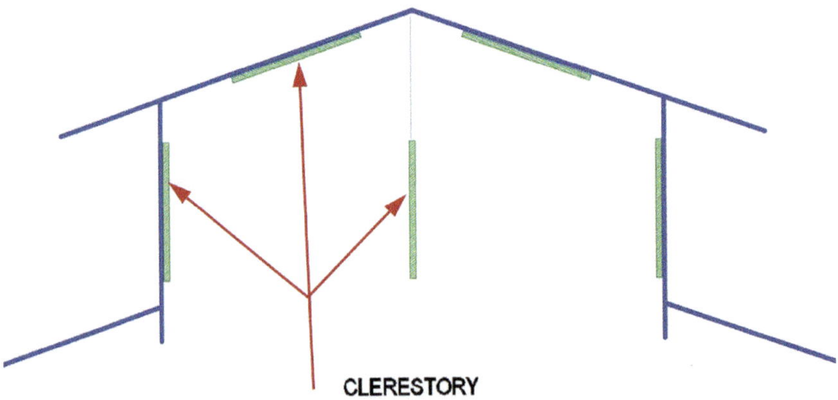

Figure 8.11 The most cost-effective clerestory treatment is to install fabric-wrapped, semi-rigid fiberglass panels along the full length of the clerestory zone in one of these three ways: (1) attached between windows, (2) directly attached to the ceiling, or (3) hung as baffles from the center ridge beam. Providing sufficient square footage of material is installed, each option will soak up some or most of the excess reverberation that builds up in this area.

8.19 Domes

A. Domes may look creative and artistic or convey a desired sense of grandeur, but they will almost always result in a room that doesn't work well for good sound creation or propagation. Domes (see Figure 8.12) have many of the same issues as spherical and round rooms.

1. Picture a room or venue with a large domed ceiling or even a smaller area with a domed ceiling that covers sound sources (loudspeakers, voices, and/or instruments) and/or an audience area. Again, without the addition of custom acoustical treatments—many of which may be generally unfeasible, aesthetically unacceptable, and very expensive to implement—domed ceilings should be avoided.

B. If you must deliver a dome ceiling of any significant size or location, be sure to include an acoustical consultant on your engineering team. Also include a budgetary line item for some *creative*—and often expensive—acoustical treatments within and/or below the envelope of the dome.

Figure 8.12 The domed, mosaic tile ceiling at St. John Neumann Catholic Church. *Source*: sjnaustin.org

1. Though not a total solution, the floor area directly below a dome must have a soft, VB-type absorptive finish. The combination of carpet and soft padded chairs/pews is a good option.

8.20 Elevations

A. Last—but far from least important—are the room's overall interior elevations.

B. If only one architectural wish could be granted, I would opt for more usable ceiling height. Here, the term *usable* means unobstructed. No other single structural parameter makes the presentation of good sound, video, and lighting more difficult and expensive than the lack of adequate vertical space, especially above and around a platform or stage.

C. The larger the room, the more ceiling height that is needed.

1. If good results are desired and/or expected, a 20′ ceiling should be considered a minimum guideline for a 5,000 ft^2 room that is being designed for audio presentations or performances. At 10,000 square feet, the usable ceiling should be at least 25′ at the lowest point.

2. Yes, of course, there are exceptions. Exceptions become workable solutions when careful, creative planning and a little greater expense are allowed to coexist.

D. For more examples of how various ceiling profiles impact acoustics, see this *AcousticBulletin* newsletter article from Ecophon titled, *The Impact of Ceiling Profiles on Room Acoustics.*[3]

8.21 Low ceilings and soffits

A. Alone, low ceilings don't render poor acoustics. However, the lower the ceiling, the greater the restrictions and limitations

3. https://www.acousticbulletin.com/the-impact-of-ceiling-profiles-on-room-acoustics/

become for good sound propagation from point source and line array loudspeaker systems.

B. Try to avoid adding any type of soffit that runs directly above a stage and/or primary seating area(s). Even a small soffit, running above any key areas, can effectively define the lowest point for the entire room or platform.

1. A mid-room soffit can easily translate into more complex and expensive audio, video, and lighting (AVL) systems.

   a. For example, if a room has a general ceiling height of 28′ but has a soffit running through the middle of the room that finishes at 24′ above the finished floor, effectively—for some or all AVL propagation purposes—the room has a ceiling height of 24′.

   b. Without adding additional loudspeakers, middle and high frequencies may be blocked or shaded and not able to reach very far beyond the soffit, loudspeakers may need to be mounted too low, camera angles and video projection cones may be clipped, and theatrical lighting may have to be mounted too low, restricting good lighting angles. See more on acoustic shadowing in Chapter 9.

   c. The mid-room soffit scenario does not necessarily impact the room acoustics as much as it impacts unobstructed AVL propagation.

2. If soffits can be pushed out to the far sides or rear edges of a room, this concern is significantly minimized.

8.22 As this commentary is being written, the implementation of AI is becoming pervasive. A Global Acoustics Ambassador at Rockfon, Pascal Van Dort, recently posted this graphic (see Figure 8.13) on LinkedIn. He asked his AI app to create an image of an open-plan office with excellent room acoustics in the style of ten internationally known architects. Figure 8.13 is my favorite of the ten. The other nine can be seen here.[4] The human imagination should be no less wondrous.

Figure 8.13 Notice that many of the themes outlined in Chapter 8 are incorporated in this rendering. Lots of complex geometry, convex features, and scattering. Even the concave elements provide some high frequency scattering. Ironically, even AI struggles with some basic acoustical concepts. There appears to be a preponderance of concave, focusing geometry shown, too, and it doesn't appear there is any absorption between the ridges.

4. https://www.linkedin.com/posts/pascalvandort_open-plan-office-with-excellent-acoustics-ugcPost-7120394764465659904-mjiY?utm_source=share&utm_medium=member_desktop

CHAPTER 9

ACOUSTIC SHADOWING

9.1 Acoustic shadowing occurs when objects block the direct flight path of sound energy between a sound source and one or more listeners. This subject comes into play more often than might be imagined. The most common examples are seen in large assembly areas that have columns, balconies, truss beams, and/or large lighting pendants or chandeliers.

9.2 This scenario is easily demonstrated and understood using this simple test: If a listener cannot see all of the nearest loudspeaker or other sound sources they are expecting to clearly hear, some or all the mid and high frequency content will be obstructed—creating an acoustic shadow. This matters greatly if your client is concerned about speech intelligibility because the primary frequencies needed to deliver good speech clarity and intelligibility fall between 500 hertz (Hz) and 4 kilohertz (kHz). Unfortunately, most of those frequencies are easily shadowed or blocked.

9.3 The trig calculations[1] needed to quantify the problem are too complex for this commentary, therefore let's stick with these basic points:

A. The largest dimension of an obstruction is not necessarily the most *significant* dimension. The most significant dimension is usually the one obstructing sound from getting to one or more listeners on the horizontal plane (see Figure 9.1).

1. https://www.prosoundweb.com/acoustic-shadowing-unpacking-wave-numbers-k-introducing-the-wave-ratio-thesis/

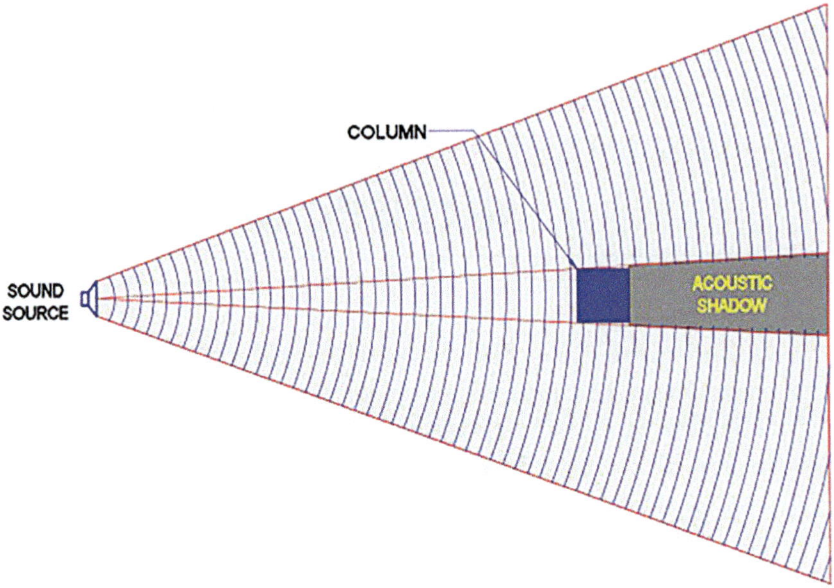

Figure 9.1 A simplified plan view of what an acoustic shadow looks like. They are much like lighting shadows but with less sharply defined edges.

1. For example, for any given seat, the significant dimension of a 12″ square column that is 15′ tall is almost always 12″, not 15′.

9.4 Acoustic shadowing can be parsed into three ranges: full blockage, partial blockage, and no blockage.

A. Full blockage: If the most significant dimension of the blocking structure is equal to or greater than two times the wavelength dimension of concern, most or all sound at that frequency—and above—will be blocked.

1. An obvious example of this occurs when a balcony overhang blocks the direct flight of sound from a loudspeaker system (see Figure 9.2).

B. Partial blockage (technically known as the *diffraction zone*): When the most significant dimension is between one-quarter

Figure 9.2 Notice that in this church sanctuary—because of the ever-expanding circumference of the pew rows—a significant amount of the first-floor seating capacity is under a balcony. Without the under-balcony fill speakers [A], many people would be in an acoustic shadow because the main loudspeakers [B] are installed well above the bottom of the overhang. *Source*: Bel Air Church

and two times the wavelength dimension of concern, those frequencies will be partially blocked or diffracted.

1. The diffraction zone includes those wavelengths that can partially wrap around the obstruction in question but without full power and fidelity. These frequencies may be audible but will be distorted—in time and energy—relative to the direct, unblocked sound arriving in the same general area.

C. Full and partial blockage examples using 1,500 Hz: 1,500 Hz is near the middle of the four octaves of sound that are most important for speech intelligibility.

1. The wavelength of 1,500 Hz is 9″. The quarter wavelength is 2.25″. Therefore, based on these dimensions, any obstruction that is 18″ or larger will completely block sounds that include content at 1,500 Hz and higher.

2. Partial blockage will also begin at that frequency and above if the obstruction is as small as 2.25″.

D. No blockage: When the most significant dimension of an impediment is smaller than one-quarter the size of the wavelength dimensions in question, no significant blockage occurs.

E. The grand takeaway: There should never be an object that blocks the line of sight between any listener position and the sound source or loudspeaker that provides direct coverage to those positions. If you want to take a deeper dive into this subject, see this article: *Acoustic Shadowing: Unpacking Wave Numbers (k) & Introducing the Wave Ratio Thesis.*[1]

9.5 I'm reminded that at some venues, *obstructed seating* tickets are sold. What is implied is a visual obstruction to what is presented on stage. In my experience, it is never an audible obstruction. Nevertheless, if you can't clearly see (presuming sufficient lighting) the front of the speaker(s) that is supposed to be providing sound to your area because of some structural obstacle you, too, will have an obstructed seat. Need I say more?

CHAPTER **10**

SPEECH INTELLIGIBILITY

10.1 When it comes to the good and bad of acoustics and sound reinforcement, few subjects receive more attention than speech intelligibility. Simply put, the phrase *speech intelligibility* references the average* person's ability to understand what is being said when another person is speaking their native language. (*Requires average hearing and intelligence and the absence of an unfamiliar dialect or thick accent.)

10.2 Loudness versus clarity: When talking to new clients, one of the most common complaints we hear is: "Our people can't hear the pastor (or presenter) who is preaching (or teaching)." Nine times out of ten, they can *hear* but they can't *understand* what is being said. This is a perfect illustration of loudness versus clarity. Good speech intelligibility and good clarity are pretty much one in the same, and neither requires excessive volume or loudness.

10.3 The English language is almost entirely dependent on the quality and clarity of the consonants communicated within each spoken word, not the vowels. Here's a written example. See if you can figure out this phrase using only the vowels. "_ _e _e_ _ _ _ea _ _e_ _ _ _y _ _e _ea _ _o _ e, _ _i _ e _ _e _ai_ i_ _ _ai _ _o_ _i_ue_." (See 10.4.B for the complete sentence.)

A. This next sentence shows only consonants. "Th_ q_ _ck br_wn f_x j_mps _v_r th_ l_zy d_g." With a little effort, most people can figure out most or all the words in the second sentence. (See 10.4.C for the complete sentence.)

B. When consonants are not clearly received, words are missed or confused with other words. Sometimes, this can be a significant problem. Examples of poor speech intelligibility arise quickly when two or more consonants rhyme. We've all experienced this at one time or another. The letters P, B, T, and D are easily confused, and M and N often get crossed up, as do F and S.

10.4 Speech transmission index (STI) and articulation loss of consonants (%ALcons)

A. Speech clarity and intelligibility are so important that actual objective tests have been devised and established as industry standards. Those standards are called the speech transmission index and the articulation loss of consonants. STI and %ALcons are covered more specifically in Chapter 4, Section 4.7.

B. From Section 10.3: "She sells seashells by the seashore, while the rain in Spain continues." Doesn't make any sense you say? Wasn't meant to.

C. From Section 10.3.A: "The quick brown fox jumps over the lazy dog." How did you do?

10.5 "Speech intelligibility is directly impacted by the level of background noise, the reverberation time, and the size and shape of a room."[1]

---

1. https://www.acoustic-glossary.co.uk/speech.htm#speech-intelligibility

CHAPTER 11

THREE SIMPLE, NONTECHNICAL ACOUSTICAL TESTS

11.1 There are three simple tests that can be used to quickly judge the most basic acoustical merits of a room. These tests cannot and should not replace objective testing that is done by a professional acoustician, but they will give anyone interested a rough idea of the acoustical conditions of an existing room. These tests are called: *the two-talker*, *the hand clap*, and *the room-mode analyzer*.

11.2 The two-talker test:

A. This test is used to get an initial impression of the following acoustical qualities: speech intelligibility and clarity, room noise, and critical distance (the point at which you hear equal amounts of both direct and reverberant sound). The results will potentially determine the need for a sound reinforcement system.

B. The procedure goes like this: Two people face each other, spaced about three feet apart. The beginning location is not overly significant. For this example, let's start the process near the front edge of the stage, platform, or lectern location in the room.

C. The person standing closest to the stage (person P1) faces out toward the audience area. The other person (P2) should face P1. P1 begins by speaking a short list of five to 10 random words, each only having one or two syllables. Pick words that contain the consonants P, B, D, T, M, N, F, and S. Speak at a normal, conversational speaking volume and never alter your speech volume, enunciation, or cadence.

D. Next, P2 repeats the words back to P1. Presuming P1 receives the first list back from P2 without any errors, P2 should then move away from P1 by about four paces, or about 10′. P1 always stays at his/her original position.

E. Next, P2 should deliver a new set of random words to P1, with P1 repeating the new list back to P2. If P2 receives the word list back without error, P2 moves away from P1 by four more paces.

F. Continue this process of exchanging words, with increased spacing, until either P1 or P2 begins to struggle to hear or understand the other person or makes obvious errors in the words they send back. If about one in five or two in 10 of the words are received inaccurately, you have reached the point of *poor-to-bad* intelligibility.

G. At this point, stop the test. There is a good chance you have reached a virtual boundary (see Figure 11.1) called *critical distance*. This means you are hearing an equal amount of direct and reverberant energy when the other person is talking. Critical distance is further explained in Section 11.2.I. There is no point in continuing further with the test. Make note of how far apart you are now standing. Also make note of how much of the room has yet to be evaluated.

   1. If P2 makes it all the way to the furthest seat in the room without a noticeable loss of intelligibility, it means the environment is free of excess reverberation or noise. If the room is not too reverberant or noisy, you are probably evaluating a fairly small room with reasonably good acoustics.

H. Presuming you have not reached critical distance, you may find a point at which the other person's voice is just too soft

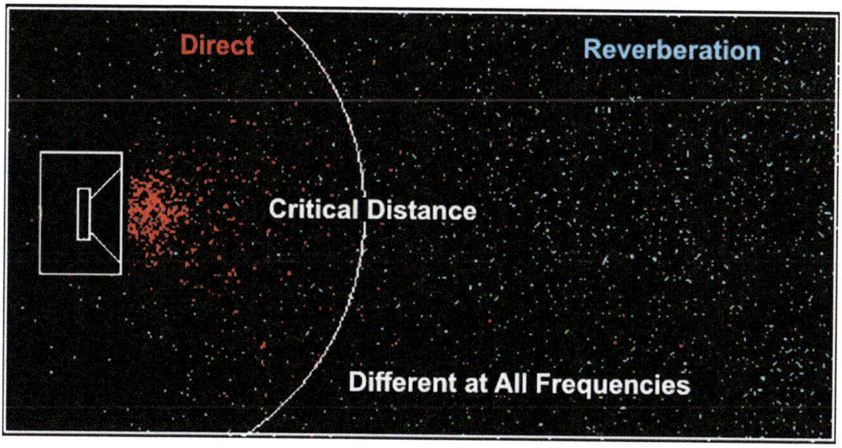

Figure 11.1 Notice the gradual change in density between the red and blue dots as the sound wave moves away from the loudspeaker. Also, be aware that the critical distance changes somewhat, based on frequency.

to be easily and clearly heard. If this happens, it indicates that a sound reinforcement system may be needed to amplify or reinforce the spoken word.

1. If the room is noisy, you may reach a point where you cannot understand the other person because the background noise is equal to or louder than their voice.

I. Critical distance is the distance from the sound source where the direct and reverberant sound energies become equal. The more reverberant the room, the closer the point of critical distance will be to the sound source (human or loudspeakers). Conversely, the more absorbent/diffusive a room is, the further the point of critical distance is from the sound source.

1. Critical distance will very likely be different at all frequencies. If one of a project's goals/deliverables is good acoustical design, the critical distance should be as far as possible from the sound source and the residual reverberation well managed and fairly even at all frequencies.

2. Direct sound from loudspeakers diminishes in level as a function of distance (inverse square law), whereas reverberation constantly spreads throughout a room. As new sound energy from loudspeakers is propagated, reverberant energy keeps building up until it becomes equal with the new incoming sound energy. When the new energy becomes equal to the amount of energy the room can absorb, the two are said to be in a *steady state* condition.

   a. Moving beyond steady state: When the reverberant energy becomes 12 decibels (dB) or greater than the direct sound, all intelligibility is lost.[1] This explains why turning up the sound system never fixes poor intelligibility or clarity; it only makes things worse.

11.3 The hand clap test:

A. This activity can be done solo. It is used to evaluate the presence of problematic echoes and provide a rough idea of a room's reverberation time (T_{mid}). The procedure for this test is no more difficult than having the ability to produce a single, loud hand clap. Slapping the fingers only (not the palm) of one hand against the palm of the other hand works best for me (see Figure 11.2).

B. Walk around the venue and occasionally stop and clap your hands together one time. Then just listen. In almost all locations, you'll likely hear either reverberation, echo(s), or both.

   1. If you hear a smooth, obvious, reverberant tail (decay) after the hand clap, without any obvious slap or flutter echo, try to *silently* count the number of seconds it takes for the reverb tail to become inaudible. Starting with zero, count—"zero Mississippi (at the point of hand clap impact), one Mississippi, two Mississippi," etc. If you get

1. https://education.lenardaudio.com/en/04_acoustics_2.html

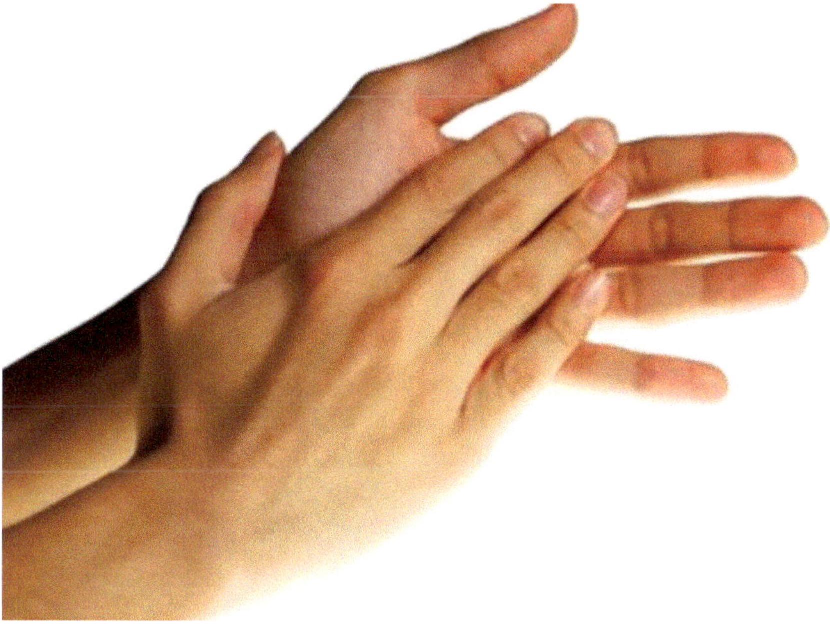

Figure 11.2 Maybe you've noticed? When it comes to applauding a performance, it seems that everyone has their own way of clapping. Clapping your hands for maximum energy and pop requires a little bit of technique and practice. Try using fingers only of one hand to strike the palm only of the other. This, and using about a 45-degree rotational offset, seems to work best for most people.

to more than about 1.5 *Mississippis* (the number + Missi = roughly one-half second), this is an indication the room probably has too much reverberation for *good to excellent* speech clarity and intelligibility. Move around the room and repeat as needed. The goal is to get a rough idea of the rooms T_{mid}, which is what most people are describing when they say a room has some number of seconds of reverberation.

2. It is fairly common to hear both echo and reverberation. When you easily hear both, it usually indicates that the T_{mid} is under about 2.0 seconds, but greater than about

1.0 second. If this is the case, treatment specifications should prioritize minimizing or eliminating the echoes.

3. It is also common for reverb to completely mask echoes—flutter or slap. When a room is treated to reduce reverberation, that treatment may *uncover* residual echoes. Keep an eye on the parallel walls when considering mounting locations for treatment. If possible, absorptive or diffusive materials should be placed on at least one surface of each pair of parallel walls/surfaces.

4. If you hear echo, with little or no reverb, look around for any large, hard, parallel surfaces you may be standing between. These echoes are symptomatic of acoustical problems that will, at the very least, reduce clarity and intelligibility.

11.4 The room-mode analyzer test:

A. This exercise requires a sound system capable of easily reproducing a very low frequency (VLF), such as 60 hertz (Hz) or 80 Hz, and a sine wave tone generator. It is used to uncloak and reveal a room's modal characteristics.

B. This test can be easily accomplished if you have access to a venue's sound system and a sound tech to help with the implementation. Ask to have a 60 Hz or 80 Hz sine wave tone run through the system. It doesn't have to be loud, although the louder the tone, the more dramatic the experience.

C. Slowly walk around the room and listen. This test is so interesting and powerful that the outcome is hard to describe. You must experience it at least once in your life. See Chapter 5, Section 5.8, for a metaphorical attempt at describing what you'll hear.

1. Next, ask to have the tone generator changed to another low frequency, such as 100 Hz. Again, walk around and listen to what happens. Every frequency will have a different modal pattern (see Figure 11.3) of peaks and

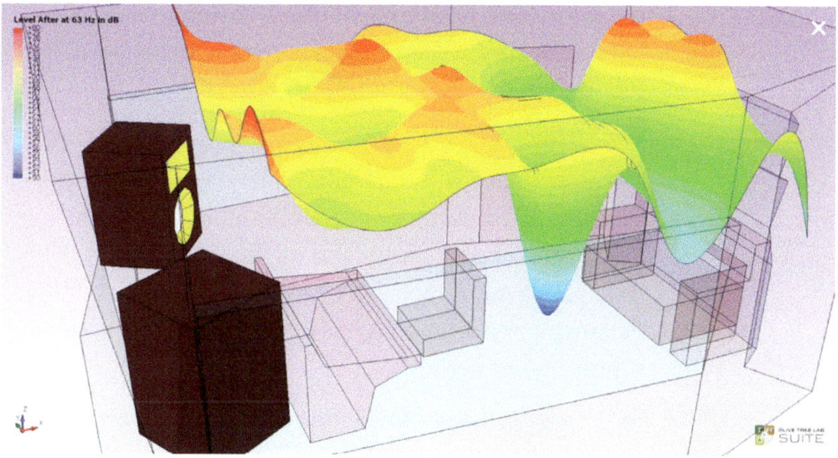

Figure 11.3 While only showing a small room, this graphic does a good job of illustrating the dramatic changes in level (volume) for a 63 Hz tone. The lime green bands represent the approximate median level, while the range, from loudest (red) to softest (blue), is 30 dB. By moving only a few feet—in any direction—a listener could easily claim they are hearing way too much or too little VLF content. *Source*: Olive Tree Labs Suite at mediterraneanacoustics.com

valleys. Geographically, this translates to peaks and valleys that change locations every time the frequency changes.

2. Next, find one location and stand or sit still while the sound tech slowly changes the frequency—up or down—in 10 Hz steps. If 10 Hz resolution is unavailable, use whatever frequencies are available between 40 Hz and 315 Hz. At the location you have chosen, each frequency change will bring a different result (change in volume). Now move to another location only two or three paces away and listen again to the same sequence of frequencies. In most rooms, the second location will vary greatly from the first.

3. The loudest frequencies in the first location may be the softest in the second. Any combination of variables in

frequency and location will produce different results. Do not be shocked when you experience this: what is too loud in one location will become near silence a few feet away. Pure audio magic!

4. The experience will be similar in most rooms, though in some, it will be more dramatic and in others less so, based on room geometry and applied acoustical treatments. In a good room, the variations in what you hear will be more subtle, which is our goal. Sadly, perfect rooms do not exist.

D. The underlying causes of this unique listening adventure are the modal nodes[2] and anti-nodes that are resonating throughout the room and changing based on the stimulus frequency, your position in the room, the room's dimensions and ratios, the architectural finishes, and the acoustical treatments applied. See Chapter 5, Section 5.5, for more on room resonance and room modes.

---

2. https://www.acoustic-glossary.co.uk/sound-waves.htm#nodes

CHAPTER 12

NOISE

12.1 The methods and materials used to minimize or contain noise are generally divided into two classes, based on the need to control or mitigate either internal or external noise. To review, noise is sound energy that is unwanted, noncomplementary, and/or distracting to the intended/desired audio content you are trying to receive.

A. Noise can either be short or long in duration. Regardless of length, it is still noise.

B. A room's ambient noise floor and its speech intelligibility scores go hand in hand.

12.2 Signal-to-noise (S/N) ratio:

A. This is a simple calculation that gauges the difference between direct, desired program material (signal) and the total, aggregate, ambient noise package in a room. Contrary to the *ratio* misnomer, the math produces a difference, not a quotient.

B. The S/N ratio is expressed using decibels (dB) as the unit of measurement. If the program material (signal) averages 85 dB sound pressure level (SPL) and the ambient noise measures 55 dB SPL, it is said that the S/N ratio is 30 dB.

C. General industry guidelines suggest that a 10 dB S/N ratio is the absolute minimum allowable for any speech comprehension. Do not infer that 10 dB is remotely acceptable, but rather, it has been shown to be the minimum value that would result in some *minimal* speech intelligibility under the most extreme emergency conditions.

D. When possible, most acousticians and audio system designers are looking for a minimum S/N ratio of 50 dB or greater.

E. While coherent early reflections and moderate reverberation are not considered noise, unintended echo, be it flutter or slap, is another form of noise.

12.3 What happens when noise and reverberation are combined?

A. Given a venue that is acoustically well matched for the program material being presented, appropriate amounts of reverberation are coherent and complementary to the desired program sounds. In excess, reverberation distracts from the quality and clarity of the direct sounds, thereby becoming much like any other ambient noise.

B. When other background noise(s) is introduced into a highly reverberant space, the S/N ratio becomes even worse.

1. All noises mix together, regardless of their source, tonality, quantity, or location of origin. When combined with and sustained by excess reverberation, these noises can create a din of incoherent sound, making even the most articulate presentation or performance indistinct.

12.4 For example (see Figure 12.1), some common structural, mechanical, and environmental noises include:

A. People talking, coughing, sneezing, clapping, laughing, crying, and/or being generally restless

B. A noisy sound system, electronic instruments, or in-room production equipment

C. Water features, fountains, and pumps—especially water that splashes

D. HVAC hum, vibration, and wind-velocity noise

E. Electrical transformer hums and lighting buzzes and hums

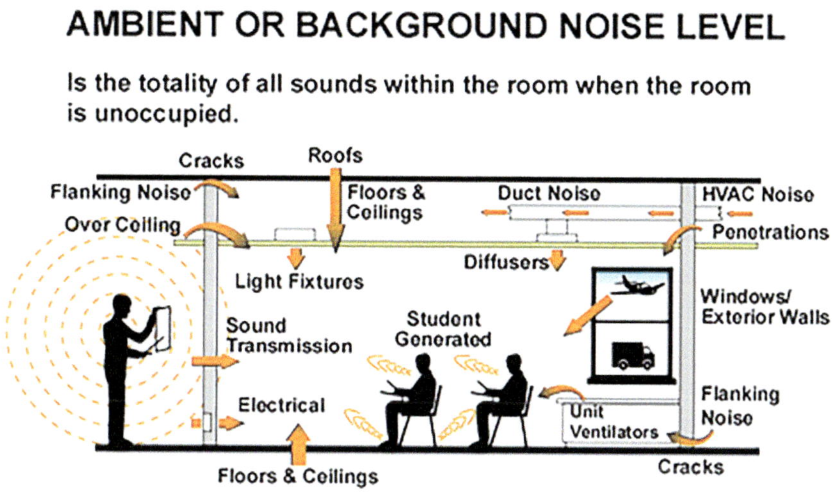

Figure 12.1 This graphic does a good job outlining the most common external noise pathways. Remember, when damping or blocking noise transmission is required, the velocity-based (VB) and pressure-based (PB) absorption treatments described in Chapter 6, Section 6.3, are ineffective tools. *Source*: AcousticalSurfaces.com

F. Elevators

G. Outside road, rail, airplane, or other environmental noise

H. Neighboring or adjoining rooms—above, below, and on all sides—that transmit loud or noisy voices, music, and/or other activities

12.5 Internal noise:

A. Noise that is created and stays within the envelope of a given room is internal noise. The first five items listed in Section 12.4 are examples of internal noise.

1. Internal noises can and should be removed or reduced at their source.

B. Unwanted noise(s) that cannot be mitigated at the source can be minimized somewhat by using absorptive VB treatments on the walls, floor, ceiling, etc.

12.6 External noise:

A. External noise that is transmitted into a room is just as problematic as internally generated noise. The last three items listed in Section 12.4 are examples of external noise.

B. Depending on the type of noise and its location of origin, blocking it from entering a building can often be more challenging and expensive than any of the internal acoustical treatments mentioned earlier. The most effective way to stop noise from getting into or out of a room is to identify the potential problem before any construction begins. At this point, an acoustical consultant can evaluate the potential problems and specify the appropriate construction materials and methods needed to block all—or much—of the unwanted sound.

1. If you are working with an existing structure, blocking noise transmission can be quite complicated. Because they are misapplied, many of the most common and obvious VB and PB acoustical treatments are ineffective and a waste of money.

12.7 Sound transmission class (STC):

A. Many construction materials have an STC rating. The STC rating defines how effective a material or assembly is at blocking various speech-range frequencies—125 hertz (Hz) to 4 kilohertz (kHz)—that might try to transmit through them via sympathetic vibration. An STC rating also factors in the mounting method required to achieve those specific results. See Chapter 4, Section 4.8, for more on STC.

B. Please memorize this sentence: *Mass blocks sound; absorptive products do not.*

1. Get to know and use the phrase *transmission loss* (TL) when discussing sound blocking. TL is the subject, not sound absorption. If TL is the goal, it takes mass to deliver results.

2. Like sound absorption, TL is frequency dependent. High frequency sounds are easiest to block, while low and very low frequencies are the most difficult.
    a. For example, an assembly that delivers 77 dB of TL at 1 kHz may only result in a TL of 46 dB at 63 Hz.
3. If you use enough mass, you can block frequencies as low as desired or needed. However, even mass has its limitations when structural vibration comes into play. In that case, mass plus structural isolation is needed for effective results.
4. Without additional mass layers, even 12″ of absorptive treatment may not keep the neighbors from complaining. If subwoofers are involved in your project, my sympathies go out to you, the contractor, the owner, and the neighbors.

CHAPTER 13

VARIABLE ACOUSTICS

13.1 The phrase variable acoustics (VA) means that the acoustical characteristics of a room can be changed to better match the requirements of a performance or presentation. This is a very useful feature.

13.2 Historically, the acoustics of most rooms have been static and unchangeable, other than when people occupy the space in varying quantities. Beginning with Wallace Clement Sabine[1] in the early twentieth century, acousticians began studying the qualities and properties of various finish materials and how they affected a venue's acoustics. It wasn't long before people started trying to manipulate and control the reverberation time in rooms.

13.3 VA is a topic that is most often considered for implementation when designing a performing arts venue, though many other facilities are candidates too. Concert halls, houses of worship, and multipurpose ballrooms are some prime examples.

A. More often than not, the goal in these performance venues is to build and maintain a space that can support the widest possible range of events and activities. In almost all cases, venues that can effectively accommodate a diverse genre of audio and musical styles are the ones booking the most calendar dates—ultimately making them the most commercially viable and profitable.

1. https://en.wikipedia.org/wiki/Wallace_Clement_Sabine

13.4 Over the past 50 years or so, concert halls and other performance venues have been installing mechanical VA systems. The most common applications and techniques included:

A. Heavy drapes, which are manually opened or closed to expose or cover various reflective surfaces. This application is *passively subtractive*, providing variable percentages of velocity-based (VB) absorption without the use of any audio electronics.

   1. Using drapes alone, the reverberation times can never be greater than what the room offers when the drapes are fully retracted (opened). Conversely, the reverberation times can never be less than what the room offers with the drapes fully deployed (closed). More modern variations use motorized drape pulley systems or other motorized assemblies.

B. The addition of purpose-built, auxiliary reverberation chambers.

   1. These chambers are empty, highly reflective annexes that share a common moveable or shuttered wall with the main venue.

   2. The shutters are manually opened or closed to introduce more or less volume into the venue's boundaries. The application is *passively additive*, as the added volume is intended to garner longer reverberation times without the use of any audio electronics. This technique has fallen out of favor since the intended results tend to be minimally effective and are usually cost and/or space prohibitive.

C. To offer a greater range of adjustment, some venues have employed both drapes and auxiliary reverberation chambers.

D. While drapes and auxiliary chambers are seen as useful advancements over most static, acoustical options, they are passive solutions that have several notable drawbacks:

1. Cost of construction
2. Loss of usable space
3. A small range of variability
4. Slow and/or labor-intensive to adjust
5. Somewhat difficult to repeat exact settings and conditions
6. None of these solutions offer the ability to optimize the T_{60} slope ratio (TSR) characteristics of the room. See Chapter 17 for an overview of the TSR thesis.

E. With these shortcomings, it is a wonder any owner would commit to the cost of this *new* technology. However, the need and demand for such acoustical flexibility are clear and justifiable when weighed against owning an unused venue that has only a single static acoustical trait.

13.5 One of the most interesting acoustical technologies available today can be described as *electronic variable acoustics* (EVA). Consider these points:

A. Apart from acoustics, all aspects of the technical systems—that is, sound, video, and lighting—in performing arts centers, concert halls, houses of worship, or other entertainment venues are adjustable via real-time human interaction. Now, we have access to tools and technologies that bring architectural acoustics under our control too.

B. In recent years, the high-level implementation of digital audio signal processing has allowed manufacturers to explore the management and manipulation of room acoustics. This is typically done using powerful microprocessors, elaborate software applications, and secondary microphones and loudspeaker systems.

1. To be clear, most of these are *active, additive systems*, meaning they add full, natural-sounding reverberation into a room to enhance the acoustical support necessary

for purely acoustic instruments and voices. They do not and cannot remove any reverberation that naturally occurs in a room. To take full advantage of such systems, a room must start with a mid-band reverberation time (T_{mid}) no greater than about one second and ideally, an *optimal* TSR grade.

2. Now, using presets, changes can be made quickly, easily, consistently, and with a much wider range of T_{mid} and TSR parameters than ever before imagined. This technology has sufficiently matured to the point where the best EVA solutions are now a viable option with few, if any, discernible, unnatural artifacts.
3. Cost will still be an issue for some because the leading brands can be quite expensive to properly implement. Contact your favorite acoustician for more info and budgetary estimates.

C. Motorized slat absorbers: The evoke 1 modular system (see Figure 13.1) from the Danish company Flex Acoustics[2] offers a motorized, subtractive form of VA. It is a modular system that is generically classified as a variable acoustic slat absorber. When the slats are open (open slots between the slats), sound energy passes into the VB absorption chamber within. The slotted perforations also provide additional low frequency (LF) and very low frequency (VLF) absorption beyond what the absorptive material would deliver on its own. By design, when *in use*, this system absorbs more LF energy than high frequency (HF) energy, thereby addressing the tendency for most rooms to have too much LF and VLF reverberation and/or resonance.

1. The slat *shims* are motorized and programmable so that each module can be opened or closed as needed in various combinations. When the slots are closed, they present a reflective surface with some modest HF scattering.

---

2. https://flexac.com/evoke-1/

Figure 13.1 The evoke 1 VA system. *Source*: FlexAcoustics

2. Once commissioned, a control panel app is provided to activate different preset open/close configurations. The app allows for real-time or scheduled settings to be activated, thereby delivering an optimal reverberation profile for a variety of musical events.

13.6 To help illustrate why VA and EVA are gradually gaining greater traction, consider these wide-ranging types and styles of music, which are best experienced when performed in a room with an optimal reverberant profile. Examples include:

A. Spoken word—Short reverb, at or below a 1.00 second T_{mid}

B. String, wind, or brass ensembles—Medium-long reverb time, between 1.50 and 2.00 seconds T_{mid}

C. Unamplified jazz ensemble—Medium length reverb time, in the range of 1.25 to 1.50 seconds T_{mid}

D. Rock, pop, blues, funk, country, hip-hop, soul, contemporary Christian, or other up-tempo, rhythmic music—Medium-short reverb, between 1.00 and 1.25 seconds T_{mid}, combined with a *good* to *optimal* TSR grade

E. Large choir, pipe organ, and/or orchestra—Medium-long to long reverb, between 1.75 to 4.00 seconds T_{mid}

F. Dramatic acting and musicals—Medium length, between 1.25 and 1.50 seconds T_{mid}

G. Without the implementation of an EVA system, there are very few, if any, venues that can support anything close to this diverse range of acoustical requirements

13.7 As we approach the second quarter of the twenty-first century, owners, architects, and acousticians are implementing a radically more ambitious approach to VA and architectural geometric expression: *shapeshifting* room volumes. The animated clip referenced here shows how the Lindemann Performing Arts Center[3] (PAC) at Brown University uses movable walls, seating, flooring, and ceiling reflectors to suit their various application and acoustical needs.

A. Initial testing of the Lindemann PAC shows the average range of T_{mid} times for five of the many possible configurations to be between 1.33 and 2.10 seconds.

B. The Perelman Performing Arts Center[4] at the World Trade Center in New York is another recent example of this new paradigm shift in architectural acoustics.

C. Remember, room volume is a key component in determining the reverberation time in any room.

13.8 Here are some basic VA and EVA takeaways for your consideration:

---

3. https://rex-ny.com/project/lindemann-brown/sectional-animation/

4. AVA consultant: Threshold Acoustics https://www.thresholdacoustics.com/

A. Old-school mechanical VA systems are expensive to integrate, labor intensive to operate, and provide only minimally effective variation.

B. Modern EVA systems are also somewhat expensive to integrate. Nevertheless, they provide a much greater range of adjustments and can be quickly and easily modified by recalling one of many presets. Presets are established during the initial EVA system commissioning.

C. Shape-shifting VA designs are also very expensive to create and require significant time and manpower to adjust.

D. There are no electronic products/solutions on the market that can reduce or eliminate excess reverberation. Therefore, to properly implement an EVA system, you must start with a room with a very short T_{mid} of one second or less. To fully optimize for all or most of the previously listed musical styles, a good to optimal TSR grade should also be a high priority.

1. Starting with a T_{mid} that is less than one second is okay. Longer is problematic because, if needed, you can never get back to a shorter T_{mid} than the venue naturally presents with the EVA system turned off.

2. If an existing room has too much reverberant energy (in other words, reverberation or modal tails that last too long), it will first require the addition of the appropriate amounts of absorptive and/or diffusive materials to achieve the desired T_{mid} baseline before an EVA system can be considered and effectively deployed.

CHAPTER 14

PSYCHOACOUSTICS

14.1 No commentary on architectural acoustics (AA) would be complete without touching on the subject of *psychoacoustics*. The human brain is such an amazing computational tool that it is easy to overlook its role in audio perception.

14.2 "Psychoacoustics is the branch of psychophysics involving the scientific study of sound perception and audiology—how the human auditory system perceives various sounds. More specifically, it is the branch of science studying the psychological responses associated with sound (including noise, speech, and music). Psychoacoustics is an interdisciplinary field including psychology, acoustics, electronic engineering, physics, biology, physiology, and computer science."[1] Though this subject has been extensively researched and documented, only a few core concepts will be expressed here.

14.3 Time is one of the most important elements related to the quality of sound perception. As highlighted in Chapter 2, Section 2.5.B, the speed of sound is extremely slow. So slow, in fact, that when listening to identical audio content, the average listener can hear timing differences between distant and nearby loudspeakers if they are offset, front-to-back, by as little as 34 feet—in some cases, even less. As a result, time plays a critical role in the discussion of how sound is perceived.

1. https://en.wikipedia.org/wiki/Psychoacoustics

A. It takes 30 milliseconds (ms) for sound to travel about 34′—that's thirty thousandths of a second. Beyond roughly 30 ms the brain gradually loses its ability to integrate the two loudspeakers or sound sources as one; thus, we begin perceiving them as two discrete signals. If not electronically corrected in the time domain, such asynchronous signals can easily reduce audio quality, tonality, intelligibility, and clarity.

B. When the same two loudspeakers are moved from a front-to-back orientation to a side-by-side orientation, a completely different set of psychoacoustic phenomena occurs—the phantom center.

14.4 Phantom center

A. The concept of a phantom center loudspeaker (see Figure 14.1) is classic psychoacoustics in action. Most people have experienced this without knowing or understanding that it is happening. For example, set up a stereo listening environment configured as an equilateral triangle created by two speakers and your head, each separated by six feet or so. Each speaker should be facing you and positioned at the same height as your ears.

B. Next, play a monaural audio signal (such as a person talking) that propagates equally from the two speakers. An easy demonstration option is to play an AM radio news broadcast through a pair of stereo loudspeakers while you stand back a few feet and directly in between the two loudspeakers.

1. What you should notice is the talker appears to be positioned directly between the two speakers. The voice won't sound like it's coming from either of the individual speakers, but rather from a point directly in the middle.

2. Headphones and earbuds can also be used, but the phantom center image moves from being in front of you, to feeling like it's centered inside your head.

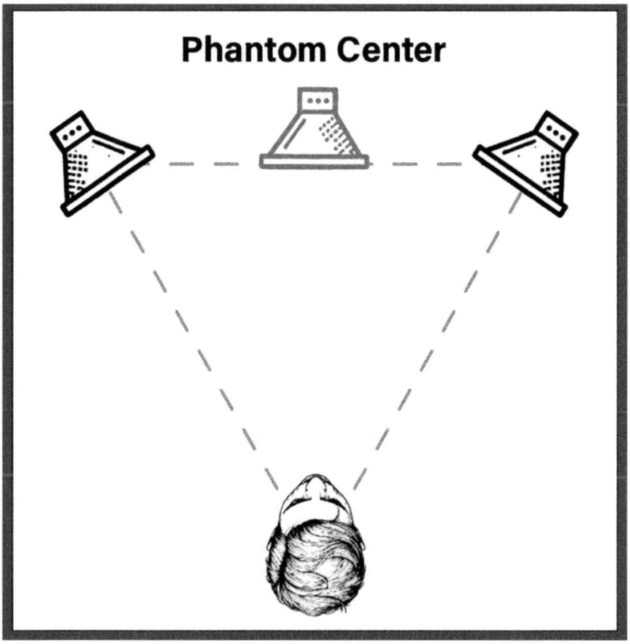

Figure 14.1 The light gray speaker in the center does not actually exist. But, when a monaural audio signal propagates equally from both the left and right speakers, it will appear to be coming out of a single speaker in the center. Graphic by Atiq Ahmed

3. Simply changing one or more of the following items on the left or right side—signal volume, timing, and/or polarity—will create a significantly different listening experience. For example, if the polarity is reversed on either of the two channels, the phantom center sound will completely disappear.

4. Because everything is equal, including the arrival time, your brain does not distinguish the sound as coming from two discrete locations. It integrates the two sounds into what seems to be a single phantom speaker, located between the two actual speakers.

5. It doesn't matter if your eyes are open or closed; the phantom center does not move or change. If you move your head slightly to one side or the other, you will begin to hear the sound move with you. Move a little to the right and the phantom center also moves a little to the right. Move too much in either direction and you may think that only the nearest speaker is on.

6. The key to this phenomenon is *time*. As you move, it is the slight differences in arrival time that make the sound appear to move with you. This is another example of psychoacoustic at work.

C. For a demonstration of time-related volume changes, take the exact same setup as previously described and add a very small amount of signal delay to one of the speakers. Let's use 10 ms, which is ten thousandths of a second. You probably won't be able to do this at home or in your car, but it is easily done with a professional sound system.

1. During this experiment, the signal will appear to be coming from the speaker that is not delayed, and the nondelayed speaker will sound slightly louder, even though the volume of each speaker remains exactly the same. This is yet another example of time-domain psychoacoustics in action.

D. There are other examples that could be used to describe the role of psychoacoustics in AA and sound reinforcement, but these should suffice for now.

14.5 How do psychoacoustics and AA fit together?

A. In a theoretically perfect venue, all listeners would sit in exactly the same optimal seat and be presented with a single, four-dimensional *sound stage* that represents all instruments, voices, loudspeakers, and complementary acoustical artifacts. There would be nothing to interfere with the sound as it travels through the air. All sounds would arrive with perfectly coherent timing, tonality, and fidelity. Being the conductor of a symphony orchestra is probably the closest anyone comes to experiencing this ideal scenario.

B. Unfortunately, this ideal listening experience is never going to happen for a paying customer in any venue. About the closest we can get to an ideal venue and listening experience is an unamplified music or vocal performance presented in a modest-sized structure with optimal acoustics. In this *best-case* scenario, each instrument and/or voice will have a direct line-of-sight pathway to each individual listener.

1. There would be no perceptible *time smear* because there is a direct one-to-one timing relationship between each instrument or voice and each listener.

2. Presuming you have a seat that is somewhat forward of the critical distance zone mentioned in Chapter 11, Section 11.2.I, everything other than the direct sound would be secondary and complementary—either integrated early reflections or moderate reverberation.

C. Once microphones, audio mixers, signal processors, and loudspeakers are brought into the equation, two or more time-dependent pathways are created between the individual instruments and/or voices and each listener. As a result, there are many more temporal variables and potential psychoacoustic distortions. If left uncorrected, these cumulative timing errors translate into uneven tonality and sound coverage, which further diminishes quality, clarity, and intelligibility.

D. Reasonable solutions become possible when architects work with the purpose of distributing chairs and seating areas evenly and symmetrically throughout a venue.

E. Audio system designers must also be allowed to place loudspeakers in very specific locations so they can present everyone with the best possible blending of direct and amplified sound; all while attempting to overcome the many timing, acoustical, structural, spatial, tonal, budgetary, and aesthetic challenges that are ever present.

1. Today's sound system designers have many tools to minimize or eliminate most of these temporal distortions, providing a room is built with good symmetry and acoustics. When asymmetrical room layouts come into play, they complicate and/or exaggerate the timing variables, in some cases, beyond what can be compensated for using today's sophisticated digital technology.

CHAPTER 15

ACOUSTICAL PRAGMATISM

15.1 Very few venues are built with *optimal acoustics* at the top of the priority list. As much as we may want it to be, architectural acoustics (AA) is not an exact discipline. It is a blending of objective and subjective goals and realities—the melding of art and science. For this reason, it is helpful to understand how to evaluate and prioritize the various opportunities and concessions that might be presented. Every project will have opportunities and compromises to be explored.

A. Because there are so many priorities and variables surrounding every construction project, pragmatic acoustical consultants are generally happy if they can get 60 to 80% of their recommendations fully implemented. Ultimately, the question becomes, what is the right 60 to 80%? An important part of the consultant's job is helping architects and owners weigh the cost-to-value of each treatment being considered.

15.2 The following sections will explore a range of categories that will provide architects with acoustically friendly solutions to consider and integrate into their designs. These are common prescriptions that should not overly tax the acoustical treatment budget, the venue's functionality, or the architectural aesthetics. Please understand that these are not complete or customized solutions, just guidelines that, in most cases, will do more good than harm.

A. Flooring

1. Carpet is often the only solution for large area, mid to high frequency velocity-based (VB) absorption.

a. A room without carpet can be challenging if there are not enough additional soft materials to help offset its absence. If carpet is not an option, anticipate that appropriate ceiling treatments will very likely be needed to provide sufficient VB absorption.

b. In many cases, installing carpet in aisles and entryways and on presentation platforms is better than having none.

c. Very few other flooring materials provide any VB absorption while also delivering the price, performance, and aesthetic values of commercial-grade carpet.

d. If AA is the only consideration, adding under-carpet padding is seldom worth the added expense.

B. Chairs

1. In most venues, incorporating padded or upholstered chairs/pews is a much better option than carpet—provided enough are used and they are permanently installed. In case this was not made clear earlier, first and foremost, effective sound absorption is a square-footage issue. The challenge is identifying how to provide enough absorptive material to make a difference.

2. Fully padded or upholstered chairs (fabric, not vinyl or leather) will provide many square feet of soft, sound-absorbing material. This is a great place to start managing the direct-to-reverberant characteristics (direct-to-reflected ratio) in any room, regardless of size.

3. Fully padded or upholstered chairs (see Figure 15.1) go a long way toward making the mid-band reverberation time (T_{mid}) more consistent, regardless of attendance. This is a key point to remember and a very helpful byproduct of upholstered seating.

Figure 15.1 Note the fully padded pews in the St. Callistus Catholic Church sanctuary in Garden Grove, CA. Even with an etched concrete floor, the acoustical character of this room was wonderful and worked well for both traditional and contemporary worship, regardless of attendance.

a. Presuming that good speech intelligibility is a priority, any room that has wide-ranging reverberation times from event to event is going to be difficult to manage. In other words, speech intelligibility will vary greatly.

b. Like it or not, attendance becomes an undesirable form of passive variable acoustics. The culprit is a T_{mid} that dramatically changes based on the number of people in attendance. See Chapter 13, Section 13.4, for more on passive variable acoustics.

c. People absorb sound. A fully padded chair absorbs about the same amount of sound as does a fully clothed adult human. When people sit in a chair, they become a nearly direct substitute for the upholstered chair they occupy.

d. When people leave, the chairs become a reasonable substitute for the people. This scheme plays out, so the room's T_{mid} has minimal variation, regardless of the number of people in attendance. This is a very desirable arrangement.

4. When fully padded seats are not an option, partially padded chairs are better than unpadded chairs.

   a. Upholstery on the back of a chair is slightly better than upholstery on the seat if there are loudspeakers broadcasting from a platform area or from some point near the front of a room. Imagine the flight path of sound coming from the loudspeakers; it is likely the sound will strike the back of the chair nearly straight on while only glancing off the seat.

   b. The opposite is true if the speaker system is an overhead, distributed type with speakers that point straight down. In rooms like this, padded seats should be a priority.

   c. For houses of worship, where congregational singing is usually encouraged, one of the best approaches is a combination of upholstered chairs, hard flooring under the chairs, and carpet applied in the aisles, on stage, and in other non-seating areas. The hard floors help support the group singing, while the padded chairs and carpet help to further manage the overall T_{mid}.

C. Walls

1. Much has already been said about wall geometry in Chapter 8. Here are a few additional guidelines relating to wall construction, materials, and techniques:

   a. A metal stud wall will absorb slightly more low frequency (LF) and very low frequency (VLF) energy than a wood stud wall.

   b. A 24″ on-center (OC) wall will absorb slightly more LF and VLF energy than a 16″ OC wall.

   c. A single-layer sheetrock wall will absorb more LF and VLF energy than a double-layer sheetrock wall.

2. Mounting sound-absorbing fiberglass panels on fir strips so they stand 1 to 2″ off the wall is more effective than if the same panel is mounted flush on the wall.

3. Unpainted brick or concrete masonry unit cinder blocks will deliver better sonic results than painted brick, painted cinder block, marble, tile, or slate. The porous surface provides surprisingly better high frequency absorption and, possibly, a little scattering if the surface is rough enough.

4. When introducing nonstandard angled geometry, there are minimum values to consider. Offsets that create too small an angular variation will deliver no appreciable performance benefit.

   a. As a guideline, 1:12 slopes or angles are the absolute minimum, while 2:12 slopes or greater are much more effective. More is almost always better, unless the angle becomes too acute.

   b. Try to work with obtuse, interior angles whenever possible, and try to limit acute angles to about 70°. Such angular offsets will tend to spread sound rather than focus it, thereby softening reflections.

   c. Mirror imaged walls, which are angled at a rate of 2:12, will produce the effective results of a single 4:12 slope angle. This can be very helpful.

d. All ratio-based slope guidelines are limited to the wavelength sizes they can affect. Larger ratios (2:12 being larger than 1:12) equal better performance at lower frequencies.

e. Angled surfaces must also be sufficiently large—at least two times the dimensions of the wavelengths needing to be managed—so they are not *invisible* to the longer wavelengths. Because of diffraction, long wavelengths easily wrap around small objects and barriers.

5. *Convex* planar curves will almost always provide softer, more pleasing reflections than *concave* planar curves, which naturally focus and amplify reflected sound. Amplifying reflected sound is generally undesirable and should be avoided in most situations.

6. Perforated (perf) paneling, mesh sheeting, and/or spaced slats can present a nice sonic and aesthetic character to a room. If such finish materials are to be used, provide a minimum of 2″ of space behind such panels or slats. If available, more space is usually better.

   a. Batt insulation or 3 pounds per cubic foot (pcf) duct liner works well in this application.

   b. If scattering or diffusion is the goal—rather than absorption—the extra space is not required.

   c. The slats and perf panels can be made of almost any rigid material and can be painted or covered with acoustical fabric.

   d. Warning—do not paint any micro-perf products. Paint must not fill any holes. If needed, it is best to have the manufacturer build the product with an acceptable color match.

e. Any added shape is also a plus. Perf panels and/or slats are also a great treatment option when concave planes are required.

7. Wall coverings, such as paintings, drapes, tapestries, posters, etc., will all have different acoustical properties. Except for the drapes, most will have minimal acoustical impact—some positive, some negative.

D. Windows

1. Acoustically, windows are difficult to treat. This is because there are a very limited number of acoustical materials that can be used without blocking a window's inherent transparency.

   a. Some promising translucent and transparent window treatments (see Figure 15.2) are noted in Chapter 16.

2. When fully closed, almost any type of drape, blind, or mechanical shade material can be somewhat helpful in reducing the specular reflections that bounce off windows. However, anything other than a soft fabric drape is virtually useless as a VB absorber. When fully *open*, all window coverings are *minimally* effective.

3. Direct application window tinting is not at all helpful as an acoustical treatment.

4. Generally, windows located above head height will present fewer acoustical problems than windows that are located at or near head height. Acoustically, *head height* ranges from about 3 to 7′ above the finished floor.

5. It is very helpful if windows are angled so they are not perfectly parallel to other windows or walls on the opposite side of the room. The direction of the angle is not too critical; however, the amount of angular offset is essential. See Section 15.2.C.4 to review the specifics.

Figure 15.2 Until recently, glass has been acoustically problematic. If your client insists on a meeting room with a large glass wall, this image shows what is currently possible using clear plastic, micro-perf, absorption sheeting. *Source*: RealAcoustix.com

E. Ceilings

1. Beyond what has been previously outlined (see Chapter 8, Section 8.17) regarding ceiling geometry and symmetry, ceiling finishes can also play an important role.

2. When it comes to steel roof decks—unless you are specifying an acoustical, perf-metal deck with fiberfill (which is highly recommended)—the bare metal roof deck will provide little or no sound absorption.

a. If left untreated, a metal roof deck is likely to require other post-construction acoustical treatment.

b. Commercial buildings that have a thin roof structure generally require some form of thermal insulation. This *thermal* insulation can be made from vastly different materials—some acoustically helpful, some not so much.

c. It can be very helpful to allow your acoustical consultant to point out the best material for the job, and this input will most likely come with little or no negative impact on the thermal performance, aesthetics, or construction cost.

3. T-grid ceiling tiles are now available in a wide variety of artistic colors, shapes, and finishes. If cost is not the only deciding factor, there may be some benefits to going with a more specialized product. See examples noted in Chapter 16, Sections 16.4.T, U, V, and W. Many other vendors listed in Chapter 16 also offer unique and specialized T-grid tiles.

4. Open beam/truss ceilings offer many opportunities for the attachment of acoustical treatment, and though usually requiring careful coordination with other building systems, it is valuable real estate that acousticians love to take advantage of whenever possible.

a. With floor and wall space often being limited and aesthetically sensitive, open beam/truss ceilings often present the best opportunity to install many of the acoustical tools detailed in Chapter 6.

F. Plants and flowers

1. Plants and flowers provide almost no acoustical value or benefit.

G. People

1. There is one unique and often overlooked acoustical fact; the quality and clarity of indoor sound can change dramatically, for better or worse, based on the number of people in attendance. If this fact is just now hitting home, please review Section 15.2.B.3.

    a. Based solely on the attendance head count, there is rarely anything beneficial about a room that changes its T_{mid} by a quarter to half a second. Such time differences may seem trivial, but they are not. This is an important issue that owners and architects can get direct control over if padded seating is allowed to be specified and installed.

15.3 Commissioning and training

A. Audio, video, and acoustical (AVA) system testing and commissioning is the process of verifying and documenting that all systems are not only working, but working optimally. Commissioning those systems is only properly finished after the carpet/flooring, permanent seating, miscellaneous furnishings, drapes or other window coverings, and HVAC balancing are *complete*! In other words, after everything else is in place and settled. Why? Because every finish detail and noise play a part in the evaluation and performance of AVA systems.

1. To properly implement the audio system balancing, timing, and tuning; perform acoustical measurements and evaluation; and adjust video focus, brightness, and contrast settings, the static ambient environment of the room must be finished, set, and stable.

2. Commissioning an acoustical design means the performance specifications and expectations are field verified to be within acceptable tolerances—relative

to the original goals. This is when the T_{mid}, T_{60} slope ratio, and any other predefined values are checked and documented.

B. The architect, general contractor (GC), and/or systems integrator can and should help manage everyone's expectations by carefully outlining the late scheduling requirements of the AVA systems commissioning and operational training. Just because the GC and their subs are done with their work does not mean the facility is ready for its first public use. It is not unreasonable or unusual for these commissioning and training processes to happen after the owner has taken occupancy of the facility.

1. Depending on the size and complexity of the installed systems, allowing for 15 to 30 days or more after receiving a temporary certificate of occupancy generally provides a reasonable amount of time for very large system testing, commissioning, operational training, and some technical rehearsals or dry runs. For large and medium-sized projects, one to two weeks is not unrealistic. Small projects may require two to four days.
2. Audio and acoustical testing and commissioning must be done when the room(s) is completely closed for all other activities—usually all day and night.

C. Audio-visual (AV) system operational training should not start until the commissioning processes are successfully completed. Why? Because it is neither effective nor appropriate to train someone on a system that is not fully configured, tested, and commissioned.

1. Just like driving a car or using a computer, the AV system operator needs to have confidence that the underlying systems are designed well and operating correctly. The major components of an installed AV system are

not end-user adjustable. During the commissioning process, the main components and devices are *locked down*. Typically, only a limited number of operational controls are given to the operator or end user.

D. Whenever possible, architects should discourage the business owner from scheduling a first use or grand opening event that is too close to the initial date of occupancy.

15.4 Value engineering (VE)

A. It is not uncommon for a sound system design to be based, in part, on the knowledge and expectation that the specified acoustical treatments will be fully and properly installed. Problems arise when some or all the acoustical treatments are unexpectedly changed, misapplied, or *value engineered* (i.e., cost cut) out of a project. Unfortunately, this is not an uncommon occurrence.

B. The following three examples are of projects that did not or nearly did not end well because owners were unable to use and appreciate their new facilities. In one case, they mistakenly concluded that the problem was the sound system. It was not. In each case, the appropriate acoustical treatment was a part of the original plans and specs. When acoustical treatments are subjected to VE, the unintended consequences can be substantial and difficult to overcome. This commentator witnessed each of these examples.

1. Case one was a college performing arts theater. A near disaster was narrowly avoided. Not all *insulation* works as a sound absorber. On this project, the insulation contractor chose to use 2″ extruded polystyrene (XPS) rigid foam, where 2″, 6 to 7 pcf, semi-rigid fiberglass was specified to be installed behind a perforated metal wall finish.

   a. Insulation is insulation, right? 2″ XPS is just as good an insulator as 2″ semi-rigid fiberglass, isn't it? XPS costs about half, so who's to notice or care?

b. This reporter happened to make an unscheduled jobsite visit for other reasons and noticed that the wrong materials appeared to be staged for installation the next day. Further evidence showed that installation of the XPS foam had already begun in a few small areas.

c. To verify the concern, a quick phone call was made to the project's acoustical consultant, followed by calls to the architect and the GC. Within a few hours, the work was halted until the correct materials could be acquired and installed. Fortunately, many additional days of rework were avoided.

2. Case two was a new high school gymnasium. The problem was that a spray-on ceiling treatment (International Cellulose K-13) was under-applied by about 75%. By looking up from the floor in this large gymnasium, no one had any way of knowing there was going to be a problem, and no one took the time to check the workmanship more closely once the installation contractor said they were finished. Everyone directly involved was completely blind to the looming acoustical problem.

   a. The result: There was so much reverberation (roughly four seconds), that the basketball and volleyball coaches could not even hold practice in their new gym.

   b. The initial blame: When the construction was complete, everyone said the problem must be the fault of a poor sound system design or installation. Naïve thinking at best since the fundamental problem existed even when the sound system was off.

   c. The fix: It took a little more than a year to work through the politics, do site testing of the *as-built* conditions, spot-check the workmanship, and finally convince the responsible contractor to reapply treatment to the originally specified thickness. The problem was adequately resolved once the installation contractor finished applying the proper amount/thickness of material.

3. Case three involves a new, large, Catholic church sanctuary. A nearly invisible ceiling treatment was deleted in the middle of the construction phase without consideration for, or communication with, the acoustician or the design/build audio contractor.

    a. The problem: The interior finishes were all marble, wood, glass, and concrete. The sanctuary had a T_{mid} of about 3 seconds.

    b. The solution: Nothing was acceptable to the owner because of the projected cost of the rework and the need to otherwise preserve the architectural aesthetics. To my knowledge, the problem still exists. So sad . . . it is a beautiful building. The high-quality loudspeakers and electronics could not overcome the excess reverberation.

CHAPTER 16

MODERN MATERIALS

16.1 Many new acoustical products have been introduced in recent years. These products have been specifically developed to encourage their use in architectural settings. Some require a velocity-based (VB) absorption backing layer to realize and/or optimize their performance.

16.2 This chapter is not intended to infer or suggest that architects try to do their own acoustical design, but rather to showcase several of the myriad treatment options available. If your current or next project is a sound-critical space, you are no longer limited to boring fabric-wrapped panels, Tectum board, carpet, or generic *Armstrong-type* ceiling tiles. This chapter primarily focuses on finish aesthetics. If you see something in the upcoming information you would like to investigate further, see your friendly acoustic consultant. They should be able to easily evaluate whether a particular material or product is a suitable acoustical tool for the desired architectural acoustics (AA) results requested or specified. See Chapter 20 for more information on applied acoustics.

16.3 The URL links listed here do not represent the entire universe of manufacturers or products. However, they will introduce some of the more prominent acoustical vendors in the world, along with the huge range of sizes, shapes, materials, and performance specifications currently available. There are even building information remodeling (BIM)/Revit files for some of these treatments. Note: Some specific links may change over time, but the homepages of each company should remain. Therefore, if a link breaks, please go to that company's homepage and search for the product.

16.4 An overview of the categories and options includes:

A. *3-D* absorption panels with complex geometric shapes, patterns, and colors

1. https://www.rpgacoustic.com/brochure/

2. https://lamvin.com/keystone-sonic-design-form

B. *3-D* scattering surfacing

1. https://gage78.com/product/ (see Figure 16.1)

Figure 16.1 These are examples of what is described as "highly dimensional and extremely durable" vertical wall surfaces. Sculptured aluminum panels such as these could be applied to provide mid frequency/high frequency horizontal or vertical sound-scattering treatments.
Source: gage78.com

C. Acoustically absorptive plaster

1. https://www.spray-on.com/sonakrete/#techdata
2. https://www.spray-on.com/sonaspray/#techdata
3. https://www.pyrok.com/wp-content/uploads/2015/02/StarSilent_Brochure.pdf
4. https://www.armstrongceilings.com/commercial/en/commercial-ceilings-walls/acoustibuilt-ceiling-panels.html
5. https://www.usg.com/content/usgcom/en/products/ceilings/integrated-ceiling-systems/ensemble-acoustical-drywall-ceiling.html

D. Fiberglass substrate with acoustical plaster skim coat

1. https://www.baswana.com/baswa-products/baswa-phon
2. https://www.fellert.com/

E. Spray-on absorption materials with color and thickness options

1. https://www.spray-on.com/specs/k13-specification/

F. Bendable, semi-rigid fiberglass paneling with fabric wrap

1. https://lamvin.com/sonic-radius-panel

G. Wood absorption—micro-perforated (perf) wall panels

1. https://soundply.com
2. https://www.realacoustix.com/wood-absorption
3. https://kineticsnoise.com/room-acoustics/acoustical-wood

H. Wood print absorption—looks like wood paneling or timber beams, but isn't

1. https://lamvin.com/walls/woodprint-panel-kit
2. https://lamvin.com/sonic-timber-beam (see Figure 16.2)

Figure 16.2 These are not structural beams. They are sound absorptive blocks that are designed to look like wood beams. *Source*: Lamvin.com

I. Wood slat panels

1. https://www.gikacoustics.com/product/slatfusor-2s-wood-slat-acoustic-panel-diffuser/
2. https://www.acoustic-products.co.uk/linea-slatted-timber-wall-ceiling-acoustic-panels/products/
3. https://lamvin.com/hushwood

J. Rigid, translucent, micro-perf, absorptive panels

1. https://www.acoustic-products.co.uk/design-composite-lightweight-translucent-acoustic-panels/
2. https://moxiesurfaces.com/product/air-board-acoustic/ (see Figure 16.3)

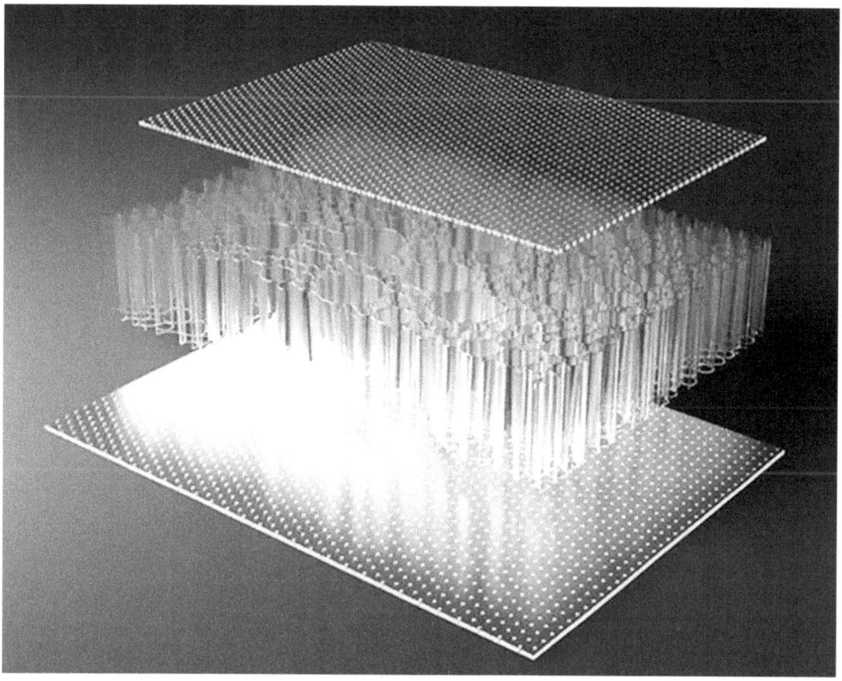

Figure 16.3 Exploded view of a translucent, micro-perf panel. This type of panel is typically specified having 1″ thicknesses. *Source*: Moxie Surfaces

K. Clear, translucent, and artistically printed films and foils

 1. https://www.rpgacoustic.com/products/?swoof=1&pa_material-composition=etfe-teflon

 2. https://www.realacoustix.com/products/glass-absorption/deamp

 3. https://www.douglas-textiles.ch/acoustic/acoustic-fabric-streamer-classic/

L. Diffusion panels with options for many geometric patterns, colors, and materials

 1. https://www.realacoustix.com/products/acoustic-diffusers

2. https://www.rpgacoustic.com/products/?swoof=1&pa_material-composition=acrylic
3. https://www.realacoustix.com/products/plastic-products/clearq

M. Concrete masonry unit absorptive blocks with a structural rating

1. https://www.soundseal.com/soundblox

N. Unique, complex, and customized perforated metals and plastics

1. https://www.mcnichols.com/perforated-metal
2. https://brankoperf.com/ (see Figure 16.4)

Figure 16.4 Perf metal is not limited to big, flat sheets. When combined with a VB backing and, if required, an acoustically transparent fabric, this type of profile provides absorption, a little scattering, and an elegant and artistic aesthetic. *Source*: brankoperf.com

O. Perforated sheetrock with multiple hole-pattern options

1. https://www.pyrok.com/wp-content/uploads/2015/02/Pyrok-Vogl-Brochure.pdf
2. https://www.gypsorb.com/

P. Digital print media on acoustical fabrics

1. https://www.acousticsfirst.com/acoustic-tone-tile-panels.htm
2. https://www.gikacoustics.com/product-category/acoustic-art-panels/
3. https://lamvin.com/sonic-printed-art-panel
4. https://whisperwalls.com/product-line/whispermural/

Q. Room dividers and screens

1. https://lamvin.com/hush-art-screen

R. Dimensional metal and wood surfaces

1. https://www.usg.com/content/usgcom/en/products/ceilings/ceiling-tiles-panels/metal-ceilings/geometrix-3-dimensional-metal-panels.html
2. https://www.imaginethis.co.uk/news/smart-home-installation (see Figure 16.5)

S. Low frequency and very low frequency absorption

1. https://www.realacoustix.com/products/bass-absorption
2. https://realtraps.com/p_megatraps.htm
3. https://www.gikacoustics.com/product-category/bass-traps/

T. Unique acoustical drop-tile ceiling treatments

1. https://acousticsfirst.com/acoustical-ceiling-treatments.htm

Figure 16.5 Just in case you thought all the other ideas and options shown are still a little boring, this scattering wall should stimulate some creative thoughts about how far you can stretch the *look* of acoustical treatments. *Source*: imaginethis.co.uk

U. Unstructured ceiling treatments (see Figure 16.6)

1. https://arktura.com/acoustic-ceiling-systems-that-achieve-both-performance-and-beauty/

V. Architectural clouds and lapendary panels

1. https://www.mbiproducts.com/products/lapendary-panels/
2. https://www.oeler.com/acoustic-products/acoustic-panels/catenary-banners/

W. Ceiling clouds that incorporate lighting

1. https://lamvin.com/zeppelin-bright

Figure 16.6 Here is an example of an intricate grid pattern for large open spaces featuring SoftGrid Flux. *Source*: Arktura.com

X. Woven wire mesh with multiple shape and color options—used as a finish over VB absorptive materials

   1. https://www.architectural-meshes.com/supplier-457226-decorative-woven-wire-mesh

Y. Weather resistant outdoor panels

   1. https://www.mbiproducts.com/products/weather-resistant-exterior-panels/
   2. https://www.sealedair.com/content/dam/protective-materials/foam/whisper/Whisper-Brochure-ENGLISH-2022.pdf

Z. Portable/modular, transparent isolation rooms

   1. https://airhush.com/#jump

AA. Septum absorbers

1. https://lamvin.com/prema/barrier-septum-absorbers

AB. BIM/Revit files

1. https://www.rpgacoustic.com/bim-files/

CHAPTER 17

NEW REVERBERATION DESIGN GOALS: THE T_{60} SLOPE RATIO THESIS

17.1 The reverberation time T_{60} slope ratio (TSR) premise: Modern loudspeakers are designed and optimized to perform in the frequency domain as flat or nearly flat audio output devices. Therefore, to optimize the listening environment and increase audience enjoyment and engagement, shouldn't architects and acousticians be designing a *good-to-optimal* reverberant timbre (in other words, nearly equal at all frequencies in the time domain) into the rooms in which these loudspeakers operate?

17.2 TSR defined:

A. The TSR thesis[1] defines specific reverberation time versus frequency goals for modern architectural acoustic environments. It is offered to advance and define the design and performance goals for sound-critical rooms and provide a simple numeric scoring scale and grade with which acoustical specifications can be initiated and/or evaluated.

B. The thesis outlines a proposed standard that condenses six octaves—63 hertz (Hz) to 2 kilohertz (kHz)—of reverberant/resonant decay-time data into a single numeric score for grading indoor performance, worship, assembly, and entertainment facilities. Specifically, it is a defining metric for scoring and grading the temporal relationship (i.e., ratio)

1. https://www.gracenoteds.com/post/2018/07/27/t60-slope-ratio

between the longest and shortest of the six T_{60} values—be they measured or predicted.

C. The T_{60} slope ratio may be represented symbolically as $T_{60}SR_6$ or TSR.

1. T_{60} is the acronym for reverberation time, in seconds.

2. SR_6 is the calculated slope ratio, using the two extreme time values—from the six octave centers between 63 Hz and 2 kHz.

3. The TSR score is calculated by dividing the longest T_{60} by the shortest T_{60}—regardless of which of the six octaves they occur. An optimal score will have a ratio between 1.11 and 1.20 (see Figure 17.1).

   a. Note: For critical listening spaces, ⅓ octave resolution is acceptable and may be desired. The same calculation (longest/shortest) is made regardless of frequency band.

Slope Ratio Score	TSR Scoring & Grading Matrix	
	≤ 1.10	Good
	1.11 - 1.20	Optimal
	1.21 - 1.30	Good
	1.31 - 1.50	Fair
	1.51 - 1.70	Poor
	Above 1.70	Bad

Figure 17.1 The TSR scoring and grading matrix. The left column represents the TSR ratio/score, not a room's T_{mid}.

4. Frequencies above 2 kHz are intentionally excluded. Because of air absorption, naturally short T_{60} values—above 2 kHz—disproportionately skew the ratio.

5. Similarly, while T_{60} measurements below 63 Hz are possible with modern measurement technology, implementing effective treatment in the 31.5 Hz octave band and below is currently an unreasonable expectation for most.

17.3 TSR scoring and grading:

A. Much like the subjective mean opinion score (MOS) that is applied to the percent articulation loss of consonants and the speech transmission index grading scales for speech intelligibility, the proposed TSR scoring and grading scale is defined by six numeric scoring tiers and five grading adjectives, which most people can easily understand.

17.4 Target goals:

A. The focus of the TSR thesis is the decay time ratio of naturally occurring reverberation and resonances found in fully enclosed, unoccupied buildings. Again, its primary application is directed toward sound critical venues, using sound reinforcement systems.

B. To achieve an optimal TSR grade, the ratio must fall between 1.11 and 1.20. At the opposite end of the spectrum, a score of 1.70 or greater indicates a bad performance grade.

C. Regardless of a venue's mid-band reverberation time (T_{mid}) goals, it is implied that the TSR guidelines remain a consistently viable metric. For example, in some very contemporary house of worship settings, a 1.00 T_{mid} might be the goal, while in a more traditional setting, a 1.50 second T_{mid} might be appropriate. Given that or most other mid-band prerequisites, the temporal TSR objectives do not change.

17.5 Modern applications: The following are some of the more obvious applications for the TSR metrics:

A. Performing arts venues: These facilities need to accommodate many disparate, artistic/entertainment productions. A good-to-optimal TSR grade will benefit most if not all musical and dramatic productions using amplified sound reinforcement.

B. Contemporary worship facilities: From the perspective of their audio, video, and acoustical requirements, these are essentially performing arts venues. They can be acoustically characterized as spaces with up-tempo rhythmic music, where one beat should not be obviously sustained much in the room while the next beat is being played.

   1. More traditional/liturgical denominations will also benefit but to a somewhat lesser degree. Fair-to-good grades may be acceptable.

C. Recording studios and sound stages: Most would benefit from an optimal TSR grade.

D. Rooms with variable acoustics (VA) and electronic variable acoustics (EVA) systems: Most would benefit from an optimal TSR grade. VA and EVA systems are detailed in Chapter 13.

E. Large rehearsal rooms: Recent European studies, such as those used to support the Norwegian Standard (NS 8178-2014), have demonstrated that rehearsal rooms should closely match performance space acoustics. Using TSR metrics for both the performance and dedicated rehearsal spaces will provide definable goals for each. For that reason alone, these venues would benefit most from a good-to-optimal TSR grade.

F. Also consider most other sound-critical indoor settings that feature both amplified music and speech.

17.6 Managing reverberation decay times is fundamentally important when evaluating rooms that have or may require a powerful, extended-range sound system. Therefore, given today's audio technology and acoustical analysis and modeling options, striving for good-to-optimal slope ratios—leading to much better music clarity—should be every bit as important as striving for good-to-excellent speech clarity and intelligibility.

CHAPTER 18

PARAMETRIC ACOUSTICS—THE PARAMETRIC METHOD OF ACOUSTIC TREATMENT

18.1 The parametric acoustics[1] thesis does not challenge the underlying physics of traditional acoustical theory. Rather, it offers a fresh perspective on how new treatment materials can and should be deployed. This technique expands on the T_{60} slope ratio (TSR) thesis outlined in Chapter 17 by providing a methodology, commentary, and examples for specifying acoustical treatments as band-limited, velocity-based and/or pressure-based absorption tools. To that end, the next level of architectural acoustical refinement is proposed: the Parametric Method of Acoustic Treatment (PMAT).

A. TSR theory prescribes a set of design and performance goals that encourage us to craft a nearly flat reverberant room response but does not offer the means or methods required to achieve those goals. PMAT adds an application methodology to this new model of architectural acoustic refinement (see Figure 18.1).

1. https://www.prosoundweb.com/parametric-acoustics-the-case-for-the-parametric-method-of-acoustic-treatment-second-edition/

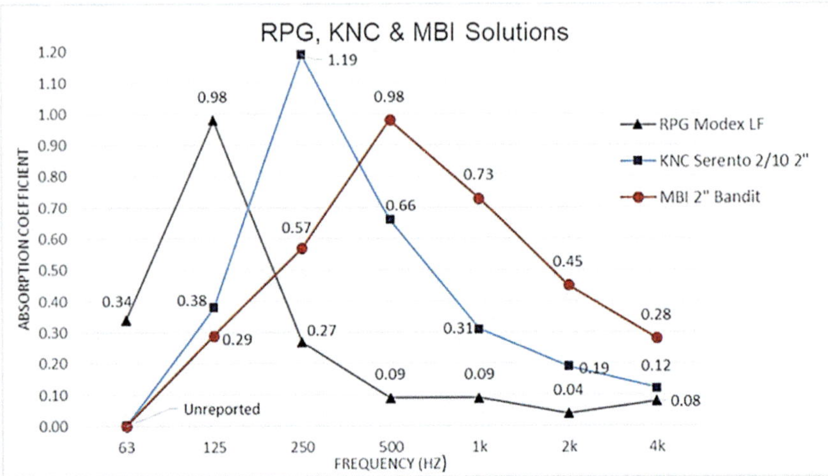

Figure 18.1 This graph shows the performance of a few good PMAT-focused materials. It features products to consider when needing to reduce reverberation in the 125 hertz (Hz), 250 Hz, and/or 500 Hz octaves. Items from RPG Acoustical Systems, Kinetics Noise Control, and MBI Products are represented here. Note the moderately narrow bandwidth of each, which is the goal for most PMAT treatment materials.

18.2 PMAT defined:

A. PMAT theory suggests that acousticians specify products and materials that perform specific work within various ⅓- to 2-octave bands. Think of these products as ⅓- to 2-octave, cut-only parametric absorbers (room-tone equalizers).

1. An electronic, parametric equalizer offers separate controls for manipulating frequency (i.e., selecting a specific center frequency within the audible spectrum), bandwidth (Q) (i.e., defining how many frequencies are being affected within a bell curve-shaped envelope surrounding the center frequency), and level (i.e., adjusting how much energy will be added or subtracted at the desired frequencies). When properly applied, these are powerful tools. If we adapt those same three parameters to material alphas, we introduce a fresh, band-specific

way of treating excess reverberation and resonance. The acoustical equivalents:

a. *Frequency*: The center frequency of maximum absorption offered by a given treatment material or product.

b. *Bandwidth*: How broad or narrow is the Q? PMAT theory suggests that when the alpha rate is reduced by 25% or more—one octave above and below its peak alpha (α)—and continues to decay at a similar rate, then the product may be well-suited for PMAT applications.

c. *Level*: Higher alphas are better. The higher the α, the more effective the product is at absorbing energy at that specific frequency. This translates to less material being needed per frequency band and, potentially, lower overall cost.

18.3 For venues offering twenty-first-century amplified music and production, acoustical designs and specifications should define appropriate mid-band reverberation time (T_{mid}) goals, while also applying customized PMAT solutions to achieve the best possible TSR score—thereby optimizing the acoustical character of the built environment.

A. While the best practitioners in architectural acoustics already understand and may implement similar principles, the PMAT concepts are currently not widely considered or implemented.

CHAPTER 19

PRIORITIES SUMMARIZED

19.1 The following is my—and my peer review team's—list of architectural acoustic priorities that architects can provide, support, and/or encourage owners to accept. All are important considerations.

A. Make good acoustics a priority.

B. Find ways to include more clear, usable, acoustically connected ceiling height. This is specifically directed to medium and large venues that rely on the simultaneous use of amplified sound reinforcement, video projection or LED walls, and theatrical lighting.

C. Encourage owners to understand and accept the value of striving for concise acoustical performance goals and, if necessary, a budgetary line item for acoustical treatments. This line item should be above and beyond the cost of the audio, video, and theatrical lighting systems and all other standard finish materials.

D. Align the base architecture with application-based acoustical goals to limit the need for acoustical band-aids after construction is complete.

E. Provide negotiable treatment opportunities and locations to manage mid-band reverberation time and T_{60} slope ratio scores.

F. Once approved during the design and construction documents phases, acoustical treatments should not be minimized or eliminated for aesthetic or budgetary reasons. It will almost always cost significantly more to retrofit acoustical treatments than it will to implement them during the initial construction phase. Further, whether specified treatments become value-engineered or eliminated, some or much of the benefits of acquiring a nice sound system will be unrealized.

G. Provide mirror image symmetry, as outlined in Chapter 8, Section 8.15.

H. Encourage and support acoustical consultants and mechanical engineers in their efforts to specify and achieve low background noise levels, that is, NC-35 or better.

I. When possible, eliminate medium and large concave or domed surfaces. If they are nonnegotiable, be prepared to address the acoustical concerns outlined in Chapter 8, Section 8.19.

J. Design all major surfaces to be nonparallel if generic or specialized acoustical treatments are unacceptable to the owner.

K. Help audio/visual system designers and acousticians manage owner expectations. Acknowledge and support the time and scheduling needs for system commissioning and training.

CHAPTER 20

APPLIED ACOUSTICS

20.1 "I imagine we might need something, but I really don't want to think or talk about acoustics." Sound familiar? It took me more than a couple of minutes to figure this out. For most venues requiring acoustical treatment, the willingness to engage and the success of such projects boils down to these four crucial feasibility questions: (1) What are the most demanding audio applications needed for the space? (2) Where can we put it? (3) What does it look like? and (4) How much does it cost? *It* being acoustical treatments.

20.2 I presume every person, business, organization, institution, and musical artist would be happy if all rooms, venues, worship centers, facilities, arenas, restaurants, and other indoor environments had *great* or even *good* acoustics. Is that really too much to ask? We certainly demand it from our lighting designers.

A. There are National Electrical Code (NEC) guidelines that define how much light each seat must receive in a building used for assembly. Sadly, there are no such formal codes or guidelines related to a room's acoustical environment.

B. Other than perhaps taste (flavor, not aesthetics), hearing seems to be the one sense receiving the least consideration by architects. Why do our listening environments garner so little respect? I think the answer is at least three-fold:

1. Acoustics is a relatively complex subject that is poorly understood by both architects and the general public.

2. Acoustical treatments must be customized for each room. There is no such thing as a one-size-fits-all solution. To reach a room's desired timbre, different sizes, shapes, locations, quantities, and finishes are required to properly address the audible frequency spectrum.

3. Acoustical treatments are generally perceived as being expensive, bulky, and/or generally unsightly.

20.3 So, how do the four feasibility questions help resolve these objections? Let's look at each question in more detail:

A. *Question 1*—What are the most demanding audio applications for the space? To begin with, there are many possible sound critical settings to consider, such as:

   1. Offices, classrooms, sports venues, performing arts centers, house of worship centers, sound stages and recording studios, libraries, restaurants, conference rooms, movie theaters, hotel ballrooms, symphonic concert halls, and hospitals, along with a variety of other multipurpose rooms.

   2. Within each of the previously listed settings, there are other specific details to consider:

      a. Is the venue meant to be as quiet as possible?

      b. Will the venue have or require a sound reinforcement system?

      c. Will the venue be primarily used for spoken word communication?

      d. Will the venue be used for the presentation of live, amplified music?

      e. Will the venue be used equally for live, amplified music and spoken word communication?

   3. This topic is covered in more detail in Chapter 5, Section 5.4.

B. *Question 2*—Where can we put it? Can architectural shapes and finishes be controlled or modified? What areas are strictly off limits?

1. The answers begin with looking at the general room geometry and all possible attachment locations found in the structure of most buildings, namely, floor, walls, ceiling, seating, windows, doors, alcoves, and soffits, etc.

2. Each of these areas presents a potential mounting/attachment opportunity. The key questions an acoustician can answer are: (1) Which areas have good viability? (2) Which areas have potential? and (3) What locations are deemed to be strictly off limits, for any reason? The answers provided by the architect and/or owner should identify all the possible locations for incorporating the necessary treatments.

C. *Question 3*—What does it look like?

1. I'm sure many have heard of or have had to work with a *pretty committee*. This group may include the architect, an interior designer, an ad hoc group in a building committee, the owner, or the owner's family.

2. There are numerous styles, methods, and techniques a creative acoustician can draw on to help satisfy the aesthetic needs of a project. There is an interesting article[1] that takes a deeper dive into the subject and presents multiple examples of nonobvious or nontraditional acoustical treatments—each example having effective results and good aesthetic appeal. See Chapter 16 for a listing of modern acoustical materials.

---

1. https://www.prosoundweb.com/parametric-acoustics-the-case-for-the-parametric-method-of-acoustic-treatment-second-edition/

3. It may be of interest to note that I've never had an attendee or paying customer complain about seeing acoustical treatment(s). Instead, it always seems to be an architect or interior designer that struggles with incorporating some new feature or shapes they didn't think of.

D. *Question 4*—How much does it cost? Obviously, there is a cost for the various treatments. Answers to the first three questions will significantly influence the fourth. But, because acoustical treatment is almost always an exercise in finding adequate locations and space, the burden of higher cost may be the only viable route to an acceptable solution.

1. For example, at the low end, expect to spend about \$5 to \$9 per ft^2 for do-it-yourself project materials. The mid-range cost could be \$10 to \$20 per ft^2 for commercially available wall panels of various sizes and shapes and finish materials. For higher-end or exotic materials and finishes, using a range of \$25 to \$45 per ft^2 should get you in the ballpark. And, of course, custom treatments will likely cost more, depending on many factors related to size, materials used, mounting options, etc. All these numbers are exclusive of shipping, tax, and installation.

2. You can probably relate to this *iron triangle*: location, appearance, and cost. You can have any two with a reasonable outcome, but the third may introduce a multiplier of significant magnitude.

20.4 As the customer, it takes very little understanding of acoustics to have a conversation that discusses those four feasibility questions from the beginning of this chapter. However, to effectively balance all four, customers need a creative, interactive design team.

A. As the architect, your willingness to engage and cooperate will be supported, affirmed, and greatly appreciated for decades to come.

B. As an acoustician, your ideas and recommendations should reveal every bit of acoustical knowledge and creativity you have to offer, while embracing the functional, aesthetic, and financial sensibilities of the owner, architect, interior designer, and end users.

C. For a deeper dive into this topic, see *Applied Acoustics: For Most Projects, Three Essential Feasibility Questions Must Be Resolved.*[2]

1. Note the title change: Since the Applied Acoustics paper was published, the *Most Demanding Audio Applications* section has been added.

20.5 Is pure random accident part of your business model?

A. When a room is built without a specific acoustical plan, its baseline acoustical characteristics are defined by its volume and geometry, the finish materials used, and its furnishings. Typically, these are materials such as: steel, wood, sheetrock, glass, carpet, drapes, blinds, padded or unpadded chairs, tile, concrete, and acoustical ceiling tiles. All are owner, architect, and/or interior designer-driven decisions.

B. While many of these materials may be loosely understood to have either *hard/reflective* or *soft/absorptive* features, the acoustical properties of each are different. The greater the total square footage for each material, the more it will influence the overall reverberation times and, thereby, the tonal character of a room. Consequently, by pure random accident, it is possible to build a venue that is too *wet* (reverberant), too *dead* or *dry* (non-reverberant), or something that is decent for some applications.

2. https://www.prosoundweb.com/applied-acoustics-for-most-projects-three-essential-feasibility-questions-must-be-resolved/

C. Structural, electrical, plumbing, fire safety, mechanical, accessibility, and energy efficiency systems all have building code compliance requirements that architects must carefully work through. So, why not apply similar time and energy probing the acoustical ramifications of your designs?

1. For new construction projects, the best way to avoid straying into a randomly accidental end product is to include an acoustician who can advise, specify, and if necessary, model the acoustical repercussions of your design choices.

2. If you believe additional acoustical treatment is required for an existing venue, it is important to properly measure the current conditions to calculate and evaluate the quality, quantity, type, and location options for each new treatment material. Using the wrong treatment type(s), installing an insufficient amount, or selecting the wrong place(s) to put new treatment(s) will likely deliver ineffective performance—potentially resulting in a huge waste of money.

20.6 Accountability: Developing a comprehensive acoustical design is an exercise in calculated probabilities. Even under ideal conditions, there are simply too many variables and unknowns for precise results. For most projects, an acoustical consultant's specifications will define treatment materials and techniques that are based on lab-tested data, published performance specifications, computer modeling, best practices, and decades of experience. From there, the important variables and unknowns include:

A. The lack of a well-defined scope of work proposal that clearly establishes the primary uses and acoustical goals for the venue.

B. The owner's and architect's willingness to accept and implement the consultant's recommendations.

C. Ill-advised value engineering.

D. Poor installation practices.

E. A margin-of-error that exists in all acoustical modeling software.

F. Because of the variables mentioned, consultants cannot and should not guarantee precise results. Look for consultants who will commit to providing a good faith effort to direct and achieve—as best as possible—the stated acoustical goals set forth in their proposals and scope-of-work documents.

20.7 Compromise should also be a part of the conversation. As noted in Chapter 15, I would rather see 60 to 80% of a design implemented, rather than 100% be rejected because any or all the feasibility issues are unresolvable.

A. At 60 to 80% implementation, noticeable improvements can be achieved and documented. From there, as-built measurements can be made that establish a baseline for any future acoustical needs.

B. Pay for a design that provides—as closely as possible—the 100% solution. Then, if necessary, let the consultant prioritize any cuts needed to meet the project's immediate goals and budget. This approach offers a reasonably clear path forward for any future treatments that may be deemed necessary.

20.8 Electronic solutions:

A. Please note that loudspeakers are not one of the subtractive acoustical tools listed anywhere in this book. Loudspeakers are additive devices; they add sound energy into a room. Most acoustical treatments are subtractive materials and devices; they capture or otherwise reduce sound energy that has been generated by loudspeakers, people, musical instruments, or other mechanical devices.

1. The best any non-EVA (electronic variable acoustics) loudspeaker can do is minimize the amount of unusable energy they broadcast into a room. This is accomplished

through the accurate selection and placement of each loudspeaker and by properly managing the timing and directivity of their sound propagation.

2. To my knowledge, there is only one electronic device on the market that is designed to cancel specific very low frequency modal energy. That product is the Bag End E-Trap.[3] Note: the application environment it is primarily designed for appears to be small rooms.

3. https://bagend.com/products/series/active-bass-trap-by-bag/

CHAPTER 21

THE *GOOD ACOUSTICS* PARADOX

21.1 When was the last time you felt or had someone tell you they visited a venue with *good* or *great* acoustics? Probably not recently, nor very often. Few people ever consider or comment on the quality of a room's acoustics. When they do, it is usually because they have a complaint about a room being too loud or too noisy, and they just *deal with* what they are being exposed to. Identifying what the noise is, where it is coming from, and why it is there in the first place are well outside their understanding.

21.2 Nationally, restaurants rank near the top of the *too-noisy* list, with the ultimate consequence being that customers will avoid returning to those establishments because the ambient noise is unbearable.

21.3 House of worship centers are another example of buildings that often exhibit less than good acoustics for music, spoken word, or both. Perhaps the room is nice for worship music but renders poor speech intelligibility. For some, the exact opposite may also be the case.

21.4 When unamplified acoustic instruments and musical vocals (i.e., choirs or soloists) perform in rooms with lots of reverberation, they can and often do sound marvelous. Why? More often than not, rooms being described by the general public as having good or great acoustics have lots of reverberation. For some unknown reason, the average person thinks that a room that has lots of reverberation equates to a room that has good acoustics. Unfortunately, there are few if any other sound sources that thrive in such environments.

21.5 The paradox: The exact same highly reverberant acoustical environment that works so well for unamplified wind, string, and brass instruments as well as for singers (not talkers) can easily become unacceptable when almost any other sound energy is propagated in the same space.

21.6 How and why is this possible? It has to do with architecturally enhanced sonic interference and time. It is the sum total of all the audio sources and noises that reach our ears versus our desire, need, and ability to receive and understand the content of the specific words and/or sounds we came to experience and, hopefully, enjoy.

21.7 Structures that are predominantly finished with hard reflective surfaces naturally initiate and sustain reverberation. Reverberation extends the length of time that all sound energy persists in a room, which is why reverb is expressed in *n* number of seconds. Desired or not, once any sound energy is introduced, it is the responsibility of the finishes, furnishings, and acoustical treatments in the room to fully capture or mitigate most of that energy so it does not interfere with the next words or musical notes that reach our ears. Short of wearing headphones or earbuds, there are no other options.

21.8 As I hope you have learned and are beginning to understand while reading through this reference, rooms having good or great acoustics are not generically defined or described as being highly reverberant spaces.

21.9 I will end with this one final note: when mostly finished and furnished with hard, reflective surfaces, small and medium-small rooms (such as many restaurants) sound loud and noisy, not reverberant. Rooms need to have a minimum volume—usually more than about 15,000 f^3—before true reverberation can develop in a space.

CODA

This brings us to the end of this tutorial. Hopefully, it has brought a new appreciation for the subject of architectural acoustics and sound and the role they play in the buildings you design and in which all of us must live and work.

Sound exists within the constraints of our primary physical dimensions of space and time. Yet when isolated to any discrete listening position, it cannot be defined by height, width, depth, mass, time, color, smell, taste, or texture. As a result, it is often underappreciated and/or ignored as something of critical importance within the three-dimensional confines of a building.

The mission of this commentary is to encourage architects to put as much time, effort, care, and awareness into optimizing the acoustics in their room designs as they put into satisfying their client's needs for proper lighting, seating, heating, and air conditioning. Please bring a renewed appreciation for acoustics into every project you assess.

Again, as Russell Johnson often said, "Acoustics is architecture!"

APPENDIX **A**

A SURVEY OF 20 ARCHITECTURAL SCHOOLS' ACOUSTICAL CURRICULA *

Arizona State University: No acoustical classes required, nor offered as an elective in the current undergrad curriculum.

Auburn University: No acoustical classes required, nor offered as an elective in the current undergrad curriculum.

Boston Architectural College: No acoustical classes required, nor offered as an elective in the current undergrad or Masters curriculum.

California Polytechnic State University, SLO: No acoustical classes required, nor offered as an elective in the current undergrad curriculum.

Columbia University, Barnard College: No acoustical classes required, nor offered as an elective in the current undergrad curriculum.

Cornell University: No acoustical classes required, nor offered as an elective in the current undergrad curriculum.

Harvard University: Does not offer an undergrad architecture program. The GSD Master's degree program offers one, 2-unit module, which briefly touches acoustical basics.

Kansas State University: No acoustical classes required, but a 3-unit elective class is found in the current undergrad curriculum.

MIT: Offers a 3-unit elective class in architectural acoustics.

---

* While not an exhaustive research project, some considerable time was spent exploring each school's website and reviewing their current architectural curriculum. 08/2024.

Notre Dame: No acoustical classes required, nor offered as an elective in the current undergrad or Masters curriculum.

Rensselaer Polytechnic Institute: The School of Architecture claims to offer the largest and most comprehensive program in architectural acoustics in North America.

Rhode Island School of Design: No acoustical classes required, nor offered as an elective in the current undergrad or Masters curriculum.

Rice University: There is one, 3-hour elective in the current curriculum.

Syracuse University: No acoustical classes required, nor offered as an elective in the current undergrad curriculum.

University of Cincinnati: No acoustical classes required, nor offered as an elective in the current undergrad or Masters curriculum.

University of Oregon: Appears to offer four, 4-credit elective classes that touch on acoustical basics.

University of Texas, Austin: One required course in general physics covering mechanics, heat, and sound.

University of Southern California: No acoustical classes required, nor offered as an elective in the current undergrad curriculum.

Virginia Tech: No acoustical classes required, nor offered as an elective in the current undergrad curriculum.

Yale University: Offers one elective architectural class that briefly touches on acoustics, with zero credits for Yale students.

INDEX

Note: Page numbers followed by *f* and *t* refer to figures and tables, respectively.